SOIL MANAGEMENT

SOIL MANAGEMENT

D. B. Davies

MA (Cantab), PhD (Newcastle)

D. J. Eagle

B Sc, MSA (Toronto)

J. B. Finney

CBE, B Sc, Dip Agric FRAg S

Farming Press

First published 1972
Fifth edition 1993

Copyright © FARMING PRESS 1972 & 1993

ISBN 0 85236 238 2

A catalogue record for this book is available
from the British Library

**Published by Farming Press Books and Videos
Wharfedale Road, Ipswich IP1 4LG
United Kingdom**

Distributed in North America by
Diamond Farm Enterprises, Box 537, Alexandria Bay, NY 13607, USA

Cover design by Andrew Thistlethwaite
Typeset by Galleon Photosetting, Ipswich
Printed in Great Britain by Butler and Tanner Ltd, Frome and London

CONTENTS

A colour section appears between pages 120 and 121

PREFACE TO THE FIFTH EDITION

Since the fourth edition of *Soil Management* was published in 1982 British agriculture has entered a less prosperous period as the European Community wrestles with the dual problems of overproduction and ever increasing costs. Two factors increasingly drive farming: the need to reduce the cost of production and the need to modify systems to meet environmental standards. The concept of Good Agricultural Practice, previously synonomous with optimum economic return, has been redefined to meet the perceived problems caused by pollution of water and air and the welfare of farm animals. The ban on straw burning, restrictions on the use of manures, fertilisers, nitrogen and pesticides are examples of the outward manifestation of these forces.

The principles of good soil management remain unchanged but some of the practice needs to be modified to meet the new requirements. Several chapters have been rewritten or revised to take *Soil Management* through the 1990s as an up-to-date text relevant to the current problems of agriculture.

D. B. DAVIES

ACKNOWLEDGEMENTS

We are grateful to the Ministry of Agriculture, Fisheries and Food for the use of results of experiments and to Experimental Centres of ADAS and commercial farms, and for information from other research organisations. Full use has been made of experience and information gained during advisory work from farms and farmers over much of the country. Wherever possible we have acknowledged sources of information in the text.

D. B. DAVIES
D. J. EAGLE
J. B. FINNEY

August 1993

CHAPTER 1

SOIL—THE FARMER'S RAW MATERIAL

Soil management has three aims: to grow crops for profit, to maintain or improve soil fertility and to avoid contaminating the environment and water supplies with nutrients or other chemicals. Sound soil management must provide a suitable medium in which seeds can germinate and roots can grow, and it must supply the nutrients necessary for crop growth. Weeds must be kept in check and the build-up of pests and diseases prevented. The traditional way of accomplishing all these aims was by crop rotation, including periods under grass, which improves the workability of the soil. Nutrients were supplied largely via the grazing animal and in the form of animal manures, while weeds were controlled by cultivations at suitable stages in the rotation.

Nowadays grass is not grown at all on many arable farms. A large proportion of the arable acreage receives no animal manures, all nutrients being applied in fertilisers. Weeds are controlled by herbicides rather than cultivations, and pests and diseases are controlled using insecticides and fungicides. The greatly increased power of tractors has made cultivation of almost any land relatively easy but their increased weight can lead to serious compaction which adversely affects drainage and root growth. Recognition during the 1960s that serious damage to soil structure was occurring led to changes in cultivation practices and later to the widespread adoption of 'tramlines' to minimise wheelings.

To appreciate the implications of all these radical changes in farming practice it is necessary to have a comprehensive understanding of the nature of soil and of crop requirements. The need for this basic understanding is all the more pressing because of the ever-increasing rate of technical advance. This short book is a straightforward account of the principles and practice of soil management which it is hoped will provide a useful background of fact for farmers, agriculturalists and others who are interested in getting the best from agricultural land.

PARENT MATERIALS OF SOIL

Soils in Britain have developed from the weathering of a wide variety of materials ranging greatly in age from coastal soils whose age can be measured in tens or hundreds of years to ancient rocks millions of years old. Most of Britain from the northern tip down to about the latitude of London has been covered by glaciers at some time during a succession of ice ages, the most recent of which was about ten thousand years ago. These glaciations had profound effects, transporting ground rock debris scoured by the moving ice from the valleys of the north and depositing it further south. The debris was either dropped where the ice melted in deposits known as 'boulder clay' or it was washed away by the melting ice. Fast-moving water carried with it gravel, sand and silt. As it slowed down it deposited first the gravel, then the sand, and then the silt. Large rivers of melting ice gave rise to fairly uniform areas of sand or gravel while smaller streams resulted in variable areas. While this was happening, strong winds blowing over the newly deposited materials caused dust storms which deposited fine sand and silt in some areas giving rise to the productive 'brickearth' soils in parts of southern and eastern England.

In still earlier times sea levels higher than today gave rise to other formations such as the chalks, formed from the skeletons of small marine life, and stone-free clays like the London Clay built up from sediment deposited on the sea floor.

The weathering processes forming soils from the parent materials include physical effects such as shattering by frost action, and chemical change by the dissolving action of water. As well as removing the more soluble materials, chemical weathering resulted in the formation of clay from the degradation products of other primary minerals. The vegetation affects the rate of weathering in several ways. Roots can enlarge crevices in rock fragments and split them as they expand. Roots affect the porosity of the soil and consequently the amount of water moving through it, which in turn affects the speed of chemical weathering. The type of tree growth can have profound effects. Deciduous trees tend to conserve nutrients by returning them to the soil each year in the leaves, whereas coniferous trees have the opposite effect.

CONSTITUENTS OF SOIL

Many types of soil are found in Britain but all contain similar basic ingredients although in different proportions. These are mineral particles of a range of sizes (sand, silt and clay), organic matter, either living or dead and decomposed, water containing plant nutrients, and air spaces or pores. The proportions of these basic materials vary not only from soil to soil but also within the same soil. For instance, in the topsoil or plough layer there is usually much more organic matter than in the underlying subsoil. The subsoil frequently contains much more clay than the topsoil, although sometimes it contains less.

The main chemical constituents of the mineral particles in most soils are silica—the chief constituent of glass—aluminium oxide and iron oxide which together account for around 90 per cent. There are also smaller amounts of a range of other elements, several of which are plant nutrients. Chalk and limestone soils are exceptions and contain up to about 90 per cent chalk or limestone which are forms of calcium carbonate.

Except on chalk soils the main constituent of sand in British soils is usually silica. Clay consists of alumino silicates containing both aluminium and silica with varying amounts of iron. Silt, which is intermediate in size between clay and sand, contains silica, alumino silicates and fragments of other minerals except on chalk soils where it is mainly chalk.

THE SOIL PROFILE

The soil profile is the sequence of soil layers which are exposed when a pit is dug to a depth of about one metre. Examination of the profile provides a wealth of information to the skilled observer.

The colour of the various layers or 'horizons' shows whether the soil is well or inadequately drained. If each horizon is brightly coloured without any mottling then the soil is well drained. Dull colours and mottling indicate inadequate drainage and the presence of grey colours known as 'gleying' indicates poor drainage. The relative proportion of grey to rust colours indicates how bad the drainage is, and the depth and intensity of gleying will indicate which horizon is impermeable.

Examination of the profile for the presence of fissures and

compact layers can show whether there is any barrier to movement of water and roots. If a compact layer is slowly permeable to water, then gleying usually occurs within it and above it.

The depth of soil accessible to roots is important since together with the texture it determines whether sufficient moisture will be available to crops.

SOIL TEXTURE AND STRUCTURE

Texture is undoubtedly the most important soil property. The size of soil particles varies 10,000-fold from the finest particles of clay, which cannot be seen even with the best optical microscopes, to fine gravel. The names used for these grades starting with the smallest are clay, silt, fine sand, medium sand, coarse sand, gravel and stones. The diameters of these particles range from less than 2 microns (2 millionths of a metre) for clay to above 2 centimetres (2 hundredths of a metre) for stones. One implication of the varying contents of these materials is a tremendous variation in the surface area of particles in comparable volumes of the soils containing them. This property increases as the particles get smaller and greatly affects the amount of water and nutrients which a soil can hold. Values of particle size and surface area are given in Table 1.1. The particle sizes are in microns, while the surface areas are expressed as multiples of the surface area in coarse sand.

Table 1.1   Size and surface area of soil particles

	Size (microns)	Surface area of same volume of different size fractions
Gravel	2,000–20,000	0.1–1
Coarse sand	600–2,000	1–3
Medium sand	200–600	3–10
Fine sand	60–200	10–30
Silt	2–60	30–1,000
Clay	less than 2	more than 1,000

Soil texture is the name given to the mixture of these particles present in a particular soil. Soil texture in the UK is classified using a system described in MAFF leaflet 895 'Soil Texture'. Soils which contain considerable amounts of all three particles are classified as 'loams', for example 'sandy loam', 'silt loam' or 'clay loam' depending on which particle is dominant. The other loams in the classification are 'sandy silt loam', 'sandy clay loam' and 'silty clay loam'. Very sandy soils are grouped into 'sand' or 'loamy sand'. Clays are classified as 'clay', 'sandy clay' or 'silty clay'. In peaty areas soils are classified as 'peat', 'sandy peat', 'loamy peat' or 'peaty loam'. When most of the peat has degraded but the soil is still relatively dark in colour the soils are classified as 'organic', for example 'organic silty clay loam'. The texture of a soil can be determined accurately in the laboratory by particle size analysis; it can also be assessed by rubbing moist soil between finger and thumb. Major classes are recognised which accurately reflect the farming properties of the soil—sandy soils, loams, silts, clays, and organic or peaty soils. Sand is recognised by its grittiness, silt by its smooth silty feel, and clay by its polish. The organic matter content affects the texture of the topsoil by making sands and clays more 'loamy', by giving more body to the sands and making the clays more workable.

The texture of a soil controls its drainage, water storage, working properties and suitability for different crops. The texture also plays a major part in determining the soil 'structure', which is the arrangement of individual particles into larger units or aggregates, which are also referred to as 'structures' or 'peds'. Because soil structure controls the agriculturally vital processes of water movement and root growth it has received a great deal of notice in recent years with changes in farming practice. The increasing weight and power of tractors and the abandonment of the ley on many arable farms have made soil structure more vulnerable to damage.

The importance of soil structure lies in the size and extent of the pore system between the structural units. It is the presence throughout the soil of a continuous stable system of both fine and coarse pores which is essential to fertility.

Because clay particles stick firmly together, even when wet, the presence of clay tends to give strength or stability to soil aggregates. Sand particles do not stick together or 'cohere' and the cohesion of silt particles breaks down when wet so sand and silt do not impart stability to a soil. Generally speaking, soils containing a lot of clay are stable and do not collapse or 'slake' when wet, although they will readily compact under the weight of implements. On the other hand, soils containing a lot of silt or sand are unstable and readily collapse when waterlogged.

An important property of clay is its ability to swell when wet and shrink when dry. The shrinkage of clay results in cracking in dry weather, which improves permeability. Alternate shrinkage and swelling are important processes in the formation of soil structure. This process, together with the disruption caused by frost, is vital to the formation of tilth or mould on heavy soils.

Although 'brickearth' and silt soils are unstable to water because of the presence of fine sand and silt-sized particles a proportion of fine sand and silt is very desirable since the amount of water a soil can hold is largely determined by the amount it contains. On the other hand, coarse sandy soils hold very little water and are droughty.

SOIL WATER AND AIR

On a hot summer's day a full crop of sugar beet will remove from a hectare of land over 38,000 litres of water; large reserves of water are therefore needed in the soil to meet the crop requirements. On the other hand, if too much water is present in the soil, roots cannot develop and are frequently killed because of insufficient aeration. Therefore, there has to be a balance between pore space containing air and pore space storing water. What we find in practice is that the larger pore spaces, greater than $\frac{1}{10}$ mm, take away the drainage water and are normally air filled, while the smaller ones store water for use by crops. The most fertile soils are those which drain freely but which retain large quantities of water and sustain vigorous growth even in prolonged droughts. Deep loams, for example clay loams, usually fall into this category.

SOIL ORGANIC MATTER

All soils are teeming with living organisms, the majority far too small to see. It has been estimated that in each gram of soil there are approximately one thousand million bacteria. When plant residues or FYM are turned into the soil they are subjected to rapid bacterial attack and much of the organic matter is lost to the air as carbon dioxide. However, a residue of dark-coloured, less decomposable material remains which is called 'humus' or 'soil organic matter'. The amount of organic matter in a soil depends on several factors, the most important being the supply of air. When aeration is restricted, due to poor drainage, then breakdown of plant remains by bacteria and other soil organisms is inhibited, so wet soils are normally rich in organic matter and well-aerated soils contain much less. The quantity of organic material returned to soil in the forms of roots, other plant remains and manures also has some effect. The actual amount of humus is the net result of input of organic materials and rate of breakdown.

The effects of organic matter on farming, discussed fully in Chapter 14, are far greater than its small amount in the soil would suggest. As well as containing a large part of the soil's reserve of nutrients soil organic matter enhances stability in unstable soils and improves the workability of heavy soils. Many of the agriculturally important processes in soil are biological in origin. For example, the transformation of organic nitrogen into nitrates, the major form of nitrogen used by plants, is achieved by several types of bacteria. The majority of living organisms in soil are harmless or beneficial but some are harmful such as some types of eelworm, foot-rot fungi, and wireworms. Although not dealt with in this book, control of these pests and diseases is a vital part of soil management.

BASE EXCHANGE CAPACITY

Soil clay and organic matter have the ability to loosely hold or adsorb a number of water soluble nutrients. This property is very important since it slows down the loss in drainage water of some nutrients, and almost completely prevents the loss of others. One type of nutrient may be replaced by another when it is added in a fertiliser so exchange of adsorbed nutrients can occur. This type of adsorption is known as 'base exchange' because the nutrients concerned are 'basic', which means they have the ability to neutralise acids. The most important base is calcium, a constituent of lime and chalk. A particular soil has the ability to adsorb a definite amount of bases known as its 'base exchange capacity'. When the exchange capacity is saturated with bases (normally mainly calcium) the soil is 'neutral'. When there is excess of lime present as in naturally lime-rich soils like the chalky boulder clays or Lias clays, the soil is basic or 'alkaline'. Soils depleted of calcium by leaching become acid.

The degree of acidity or alkalinity of a soil is very important since most crops will not tolerate strongly acid soils. Although present-day acidity can be readily corrected by liming, the occurrence of acid conditions in the past has had pronounced effects on many soils which cannot be reversed. Acid conditions intensify the effects of leaching so naturally acid soils tend to be short of nutrients. Naturally acid clays are generally much less permeable than lime rich clays and so are often poorly drained, difficult to manage and difficult to drain artificially.

NUTRIENTS

There are at least a dozen nutrient elements which are essential to plant growth. These are classified as major nutrients (Chapter 2), which are required in relatively large amounts, and minor nutrients (Chapter 3) or trace elements, which are required in very small amounts. Some nutrients, like the nitrate form of nitrogen, are soluble in water and can be completely washed out of the soil by rain. The trace element boron is less soluble and leaches more slowly while basic nutrients like potash and magnesium are loosely adsorbed and leach very slowly. The other minor elements and phosphate are strongly adsorbed. Even when applied in the form of water-soluble fertiliser, phosphate rapidly becomes insoluble after incorporation in the soil.

Generally only a small proportion of the total amount of nutrient in a soil can be readily absorbed by crop roots. The amount which can be readily taken up is described as being 'available'. When available nutrients are absorbed by a crop they are slowly replenished from a much larger pool of the same nutrient which is not readily available.

Most soils have sufficient of all or nearly all the minor nutrients required by plants. Generally, though, the major elements— nitrogen, phosphate and potash—have to be regularly applied in the form of inorganic fertiliser. Despite many claims that nutrients from manufactured fertilisers are not as good as those from organic sources, and are even deleterious to the land, there is no evidence to substantiate these statements. Indeed, organic nutrients are of no value to plants until they have been converted by bacteria into the same inorganic forms that occur in fertilisers. It would be quite impossible without manufactured fertilisers to sustain on all our land the yields necessary today. However, regular use of farmyard manure can give structural benefits by making difficult soils more workable, and it is sensible to use organic manures when they are available at low cost.

THE SOIL AS A MEDIUM FOR CROP GROWTH

In spite of a considerable amount of research it is still not possible to say in precise scientific terms what an ideal tilth should be. We cannot, for example, give a recipe for the ideal ratio of small to large clods or for the optimum amount of consolidation required; but although there is still more skill than science involved in preparing a seedbed, nevertheless a lot is known about the requirements of roots and adverse conditions can be readily recognised.

Roots require a warm, moist, friable and aerated environment. For warmth, good drainage is essential. A wet soil is cold partly because a lot of heat is needed to warm up the water it contains and partly because the continuous evaporation which takes place absorbs heat and so keeps the soil cold. Consequently, quickly draining soils like the sands warm up more quickly than heavier soils, but on the other hand the temperature fluctuation in a sandy soil is much more rapid than in a soil which holds a lot of water. During a frosty spring night sandy soils cool down most quickly and radiate least heat, so crops are more prone to frost injury.

Growing roots require oxygen and constantly give off carbon dioxide. Unless there is a continuous flow of air into the soil and carbon dioxide out oxygen becomes depleted and airless or anaerobic conditions will result. When this happens, due to compacted soil or waterlogging, not only do roots stop growing but they can be killed by toxic substances which form. Once the supply of air runs out, the soil bacteria and other organisms which require oxygen are suppressed but 'anaerobic' bacteria can flourish. One kind of anaerobic bacteria, known as denitrifying bacteria, can extract chemically combined oxygen from nitrates resulting in the loss of available nitrogen as a gas. Others can extract oxygen from sulphates forming sulphides which are poisonous to roots; other toxic chemicals are also formed. Although the whole process can be quickly reversed once aeration recommences, the check to root growth and loss of root can have a serious effect on yields if it occurs during a critical stage of growth. In cereal crops, for example, if anaerobic conditions prevent root proliferation at the tillering stage and the roots are checked until it is too late for adequate tillering then a thin stand and poor yield result.

To allow adequate aeration, drainage must be satisfactory and there must be a continuous system of pores connecting the root zone to the air above unbroken by compact layers.

Roots grow by pushing their way into pores or cracks in the soil and then expanding. Obviously the soil must be sufficiently porous and friable to allow rapid root growth but it must also be firm near the surface. If the soil around a seedling is too loose, when its roots try to push downwards they will tend to push themselves upwards and so penetration will be prevented. This is most likely to happen when a loose friable surface tilth overlies a compact layer immediately beneath. Ideally the seedbed should be a little firmer in the top 75–100 mm than below. Most farmers are aware of the need for some consolidation, but too many concentrate on getting about 50–100 mm of tilth and are often unaware of a compacted layer below which can restrict root penetration directly, or indirectly by impeding drainage and causing waterlogging.

Although most of the soil's nutrients are contained in the topsoil and the majority of plant roots remain within it, some roots must go much deeper in order to obtain sufficient water during dry spells. A root range of at least a metre is necessary in the drier eastern counties.

WORKING PROPERTIES OF SOIL

The properties of soil which affect plant growth are important but the mechanical behaviour of soil as it affects cultivation, implements and traction is just as crucial to the farmer. It is frequently the major cause of cropping limitations. At the extremes of moisture content cultivation of the soil is very difficult, if not impossible, the only exceptions being light sandy soils which are friable even when dry. The intermediate moisture range within which conditions for cultivations are good is known as the friable range. This friable range is large on clay loams, so there is a wide range of moisture content suitable for cultivation. However, on silt loams and sandy clay loams the friable range is small which means that hard clods are made sticky or 'plastic' by a comparatively small amount of rain, and conversely wet soil can quickly dry into hard clods. Consequently, these soil types are difficult to manage. In other respects they are also difficult. Weathering of a clay or clay loam results in a tilth which will not collapse when wet, but silty or sandy clay soils do not have this stability and readily run together or 'cap'. The cap is only slowly permeable to rain and is slow to dry out.

If soils are worked when the moisture content is above the

friable range and into the plastic state, they more readily compact to form impermeable layers or pans. The smearing caused by implements or spinning tractor wheels is particularly damaging since it seals off the soil's interconnecting system of pores. The seal prevents drainage and can result in the soil slaking above it to form pans several millimetres thick.

Although sandy soils are easy to manage because they are friable even when dry and do not become plastic when wet they nevertheless will compact if worked while they are wet. When this happens root restriction accentuates their droughtiness.

EFFECTS OF SOIL ON PESTS AND DISEASE

Soils and soil conditions affect the growth of crops indirectly by their effect on weed growth, pests and disease as well as directly by supplying water and nutrients to the crop.

Weeds have the same requirements for nutrients as crops but they are often more tolerant of adverse conditions. Blackgrass, for example, is more tolerant of wet conditions than arable crops. If the crop is checked by wet root zone conditions caused by soil compaction, then the blackgrass is encouraged because the crop is not able to compete. Nutrient deficiencies or acidity can have the same effect.

Adverse soil conditions due to soil compaction or poor drainage greatly increase the chances of serious infection with root fungi such as take-all in wheat and foot rots in peas and beans. If soil conditions are good and growth is unimpeded, the risk of damage by root fungi is minimised although take-all can still be a problem on friable sandy and peaty soils.

SOIL TYPE AND EELWORM ATTACKS

The occurrence of problems with nematodes (eelworms), microscopically sized pests which attack the roots of crops, is greatly affected by soil type. Nematodes such as potato cyst nematode (potato root eelworm) and cereal cyst nematode (cereal root eelworm) thrive on sandy soils and can greatly reduce yield if these crops are grown too intensively. Nematodes do not do so well on heavy soils and cereal cyst nematode is not a hazard on clays. Potato cyst nematode, however, is more tolerant and gives trouble also on heavy soils and on peats.

As nematodes attack roots their effect on crops is more serious if root growth is already restricted by adverse soil conditions.

Although poor soil conditions alone can drastically affect yield, the effect of poor soil conditions combined with an attack by a pest or disease is even more serious and can often cause crop failure. If the crop rotation followed does not control pests and disease, then it is even more important to ensure good soil conditions than in a clean rotation.

CHAPTER 2

PLANT NUTRITION
The Major Elements

The six essential plant nutrients—nitrogen, phosphorus, potassium, magnesium, calcium and sulphur—are known as major elements. The term 'major' implies that they are required by plants in relatively large amounts, not that they are more important than other nutrients. The quantities absorbed by crops amount to kilograms or tens of kilograms per hectare compared with only grams per hectare of the trace elements or minor nutrients. Although grouped together, the major elements vary greatly in their behaviour in the soil and their function within the plant. A seventh major nutrient, sodium, is essential for maximum yield of a few crops, the most important of which is sugar beet.

Nitrogen, phosphorus and potassium are the only nutrients required frequently on most land. The content of phosphorus and potassium in fertilisers is conventionally expressed on the basis of their oxides and referred to as phosphate and potash. Magnesium is required on many soil types but it is most likely to be seriously deficient in sandy soils. Sulphur is required on oilseed rape and grass in some areas.

Fertiliser recommendations are customarily given as kilograms per hectare and fertiliser bags generally contain 50 kg. The number of kilograms of nutrient contained in each bag is obtained by dividing the percentage of nutrient by two. For example a 50 kg bag of muriate of potash (60 per cent potash) contains 30 kg of potash and a bag of superphosphate (20 per cent phosphate) contains 10 kg of phosphate.

Although a small proportion of calcium is contained in many fertilisers, the amounts are insufficient for crop requirements and the main source of calcium is ground limestone or chalk.

The main sources of sodium are agricultural salt and kainit but it is also contained in a few special compounds.

13

BALANCED NUTRITION

It is important to keep a correct balance between the major nutrients. If more than one nutrient is deficient the effect of supplying only one of the nutrients may not be very beneficial and may even be harmful. By accentuating nutrient imbalance in this way the deficiency of the nutrient or nutrients not supplied is aggravated and crop yield may be depressed.

PLACEMENT OF FERTILISER

On soils giving large responses to a nutrient it is usually advantageous to place the fertiliser near the seed. Combine drilling of phosphate and potash with the seed for cereals gives larger responses than broadcasting on land which is deficient in these nutrients. In spring cereals a yield advantage of about $\frac{1}{4}$ tonne per hectare from placement of nitrogen is usually obtained on highly responsive light land, but there is a risk of delayed germination due to scorch in dry seasons. There is, however, no benefit from combine drilling on land not deficient in phosphate and potash and giving only moderate responses to nitrogen. Indeed since the rate of work tends to be slowed down, combine drilling on fertile land may result in lower yields due to late sowing.

Placement of fertiliser near the seed for potatoes gives larger responses than broadcasting on land deficient in phosphate or potash. On land of average fertility a little less fertiliser is needed to obtain optimum yield by placement than by broadcasting assuming no damage occurs due to scorch. However, if scorch occurs the maximum yield will be obtained by broadcasting the full rate. As the risk of scorch is increased by dry conditions and by use of chitted seed, placement is unlikely to give any benefit on many farms. As with combine drilling, any gain from placement can be outweighed by a delay in planting due to a slower rate of work.

SOIL AND PLANT ANALYSIS

The soil can be readily analysed for its content of available phosphorus, potassium, magnesium and calcium (pH and lime requirement) but there is no suitable method for available nitrogen or sulphur. These analyses give a guide to the soil's supply of nutrients but owing to soil variability and consequent sampling error, seasonal differences and effects of soil structure on root growth, etc., it is never possible to give fertiliser recommendations with a high degree of precision. With properly taken samples they are, though, quite capable of distinguishing deficient soils and rich soils from those with normal levels. Results for phosphorus, potassium and magnesium are usually expressed on an index basis, such as the ADAS system in which an index of 0 implies serious deficiency, while an index of 3 is about sufficient for arable crops.

Analysis of leaves is often useful in helping to diagnose major element deficiencies by comparison of leaves from deficient and normal areas in the crop. However, leaf analysis suffers from the disadvantage that nutrient contents vary continuously throughout the season, and lower leaves, for example, have different levels from upper leaves. Because of these difficulties leaf analysis cannot in general be used as a guide to fertiliser recommendations for arable crops, although it is helpful in fruit growing, using carefully standardised sampling techniques, as a guide to the following season's fertiliser needs. Sulphur deficiency in oilseed rape can be diagnosed by analysis of leaves taken at the early flowering stage. The results of analysis can be used to predict the possibility of deficiency in following years.

NITROGEN

Because it is more often deficient than the others nitrogen is the most important nutrient for most crops and for a number of reasons it is also the most difficult one to apply in the correct quantity. More than any other nutrient nitrogen controls the rate of growth of crops, the amount of leaf produced and the stage of maturity. A deficiency or excess can drastically affect crop yield. Plants short of nitrogen are pale in colour and the old leaves tend to be even paler than the young leaves. The rate of growth is poor and plants mature prematurely. When nitrogen is excessive plants are very dark green, leafy and slow to mature.

Organic Nitrogen

Most of the soil's nitrogen is contained in organic matter, both fresh and humified. There are negligible amounts in the geological materials from which soils are formed. Apart from peat soils, which consist mainly of organic matter, nearly all the soil organic matter—and hence the nitrogen—is contained in the topsoil. Every season, part of this organic nitrogen is broken down by numerous soil bacteria and other micro-organisms to the available 'inorganic' nitrate form which plants can absorb. This process takes place in two main stages. The organic nitrogen is first converted into ammonia which is then changed to nitrate by 'nitrifying' bacteria. Thus both ammonia and nitrate, constituents of ammonium nitrate, the commonest nitrogen fertiliser, occur naturally in soils.

Addition of Nitrogen to Soils

Apart from applications in fertiliser, nitrogen is added to the soil by natural processes. The most important of these is the 'fixation' of atmospheric nitrogen by bacteria in the nodules of leguminous plants such as clover, lucerne, peas and beans. Most legumes can obtain all the nitrogen they need in this way and the root residues eventually add nitrogen to the soil. Other types of bacteria also add small amounts of nitrogen. Rain contains a very small amount of nitrate formed by the action of lightning on the nitrogen and oxygen in the air, and ammonia in the air originating from industrial pollution adds some nitrogen to the soil.

Release of Inorganic Nitrogen

The amount of nitrate formed by microbial action during the growing season depends upon the quantity of soil organic matter and also on the proportion of nitrogen it contains. Both of these depend upon the previous cropping and the soil type. Soil organic matter is most readily built up by periods under grass because of its large amount of root, but the proportion of nitrogen contained in the organic matter depends greatly on the management of the sward. Grazed grass, particularly if clover is present, adds organic matter rich in nitrogen, but cut grass results in organic matter low in nitrogen. Texture also has an effect and light sandy soils generally contain less organic matter than heavy ones. Old pastures can contain up to about 6 per cent organic matter (0.3 per cent nitrogen) in low rainfall areas and up to nearly 10 per cent (0.5 per cent nitrogen) in high rainfall areas.

Under continuous arable cropping soil organic matter levels are inevitably lower than in ley arable systems because less root residue is ploughed in. Levels in old arable soils vary from about 1 per cent in sands (0.05 per cent nitrogen) to about $3\frac{1}{2}$ per cent in clays. In arable soils less nitrogen is released by soil micro-organisms, so more fertiliser nitrogen is required for optimum yield. Crops grown after straw crops generally need more nitrogen than after other crops because the stubble is low in nitrogen and high in carbon. The organisms breaking down the stubble need nitrogen as well as carbon. Because there is insufficient nitrogen in the stubble for their needs, they take it from the soil and so compete with the crop.

Nitrogen Cycle

Nitrates are soluble in water, and so can be readily absorbed by roots, but they can also be washed out of the soil by prolonged rain and lost. The processes by which nitrogen is gained and lost from the soil, is absorbed by plants and is returned to the soil in the form of organic matter are summarised in fig. 2.1.

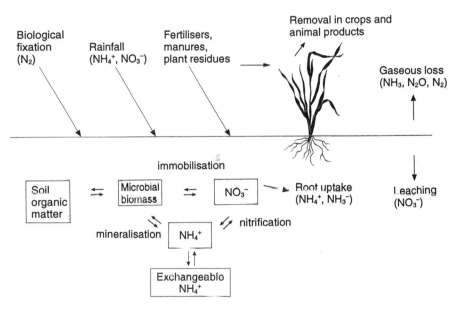

Figure 2.1   The nitrogen cycle

Leaching of Nitrate

Nitrate nitrogen is rarely leached from the soil during the growing season except on coarse sandy soils or in high rainfall areas. However, nitrate left in the soil in the autumn will be lost during the winter except on moisture-retentive soils in eastern England where appreciable amounts of nitrate remain in the soil during dry winters. Allowance must be made for these residues when deciding fertiliser rates. In the wetter parts of the country any nitrate accumulated during the summer is invariably lost during the winter, but some nitrate can build up by release from the soil organic matter when the late winter is mild and dry. This is most likely to be important on rich soils and in such seasons less fertiliser nitrogen is needed.

Effect of Season

As well as these effects of winter rainfall on the quantity of nitrogen available in soils, the weather during the growing season has marked effects on the amount of fertiliser nitrogen needed. Apart from the obvious effects of adverse weather, e.g. on ripening cereal crops, the rainfall during the early vegetative stages of growth determines to a great extent what proportion of the crop's needs are supplied by the soil and how much is needed from the bag. When there is adequate moisture available the soil supplies considerable amounts of nitrogen which are readily absorbed by roots, so nitrogen fertiliser requirements are small. In dry seasons requirements are high because less nitrogen is supplied by the soil and roots absorb their moisture mainly from the subsoil where there is little nitrogen. Nitrogen requirements on sandy soils tend to be high partly because they are droughty and partly because they are low in organic matter.

Adequate moisture reduces the need for fertiliser nitrogen, but on the other hand excessive rainfall during vegetative growth can increase fertiliser requirements by leaching nitrogen from the soil or causing it to be lost by 'denitrification'. Denitrification, the conversion of nitrate to gaseous nitrogen which is lost into the air, occurs on poorly drained or compacted soils on which waterlogging has resulted in the exclusion of air. So extra nitrogen is often needed when soil conditions are poor.

Prediction of Nitrogen Requirements

There is no completely satisfactory soil test for available nitrogen.

Recommendations on the basis of the total nitrate content of the soil determined in samples taken to depth are used in France and The Netherlands. These systems have been examined and found to be helpful in British conditions when nitrogen residues are very high. Fertiliser recommendations are made on the basis of soil type and of previous cropping and manuring in both short and long term. The amount recommended is reduced if soil analysis shows nitrate residues are high. The crop grown in the previous year can affect nitrogen requirements. Dressings of farmyard manure, legumes and grazed leys leave organic nitrogen in the soil, which becomes available to the succeeding crop, while cereals and cut grass leave less. Heavy rates of nitrogen applied to crops such as potatoes or brussels sprouts will leave some nitrate on moisture-retentive soils if the rainfall from November to February is less than about 200 mm, as it often is in parts of eastern and midland England. None will be left if it is more than this amount, if the soil is sandy and easily leached or if the soil is not well drained so that denitrification can occur.

The cropping history over the previous ten years or so also has an effect on nitrogen requirements. If leys occupy about half the rotation or frequent dressings of farmyard manure are applied, soil organic matter levels remain higher, and nitrogen requirements are lower than in all-arable systems with no muck.

A simple system of giving advice on nitrogen manuring is the 'Nitrogen Index' system devised by ADAS. This is a system of three indices, 0, 1 and 2, assigned to fields depending on cropping and manurial history, etc., as described above. An example is given in Table 2.1 for all-arable cropping systems.

Table 2.1   The nitrogen index

Previous cropping and manuring	N Index
Cereals, sugar beet, linseed, short-term cut leys	0
Oilseed rape, beans*, peas*, potatoes, short-term grazed leys with high nitrogen use	1
Large and frequent use of manures, longer-term grass with high nitrogen use	2

* Beans and peas leave N residues midway between index 0 and 1

Nitrogen Recommendations

The amount of nitrogen recommended ranges from nil to a small amount at index 2 and a high rate at index 0. The actual amount for any particular crop depends upon soil type. General recommendations for some common crops are given in Tables 2.2–2.4. More complete recommendations for a wide range of crops including grass are given in *Booklet 2191* published by HMSO. The manuring of grass is not considered in this book because the need for fertilisers on grass is decided by sward management rather than soil type and soil management.

Most of the nitrogen for winter cereals should be applied at the early stem extension stage, usually mid to late April. It is advisable to apply up to 40 kilograms per hectare of this during early tillering, usually early in March, for crops on sandy or shallow

Table 2.2 Recommended nitrogen rates in kilograms per hectare for some arable crops

	Early, seed and canning potatoes			Maincrop potatoes			Sugar beet			Carrots		
	N Index											
	0	1	2	0	1	2	0	1	2	0	1	2
Lincolnshire limestone soils	—	—	—	220	160	100	140	100	75	—	—	—
Sandy soils	180	130	80	220	160	100	100	75	50	50	20	0
Other mineral soils	180	130	80	220	160	100	100	75	50	—	—	—
Light and medium Fen silts and chalk soils	180	130	80	220	160	100	125	100	75	—	—	—
Light and loamy peats	—	—	—	130	90	50	50	25	0	0	0	0
Other peaty and organic soils	—	—	—	180	130	80	75	50	25	—	—	—

	Winter wheat*			Spring wheat			Oats, rye, spring barley			Winter barley		
	N Index											
	0	1	2	0	1	2	0	1	2	0	1	2
Sandy soils, shallow chalk and limestone soils	175	125	75	150	100	50	125	100	50	160	125	75
Fen medium and heavy silts, clays	150†	75	0	125	50	0	100	60	30	140	100	40
Other mineral soils	150†	100	40	125	75	30	100	60	30	240	100	40
Fen light and loamy peats	50	0	0	40	0	0	40	0	0	50	0	0
Other peaty and organic soils	90	45	0	70	35	0	60	30	0	90	45	0
Organic soils over chalk	90	45	0	70	35	0	70	35	0	90	45	0
Warp soils	150	100	40	125	75	30	40	0	0	140	100	40

* See Table 2.3 for mineral soil recommendations.
† Increase to 200 when disease control is good, lodging is not expected and a yield of 8 tonnes per hectare is anticipated. Apply 40 of this at early tillering.

soils, for direct drilled crops, for crops on structurally damaged soils or for thin crops.

Exceeding the recommendations can be as detrimental as applying insufficient nitrogen. By delaying maturity, yield of early potatoes, for example, may be reduced. In sugar beet yield of sugar may be lost by reduction in sugar content. Excessive nitrogen may reduce yield of cereals even if lodging does not occur, and the keeping quality of crops grown for storage is often impaired by excessive nitrogen.

Table 2.3   Recommended nitrogen rates in kilograms per hectare for winter wheat on mineral soils

	N Index				
	0		*1*		*2*
		Field beans	*Dried or lining peas*	*Potatoes, OSR High N grazed ley*	
Clays	180	155	140	100	20
Medium soils	190	170	160	130	70
Deep silts	160	130	115	70	0
Sands, shallow chalk and limestone soils	200	190	180	160	130

Table 2.4   Recommended nitrogen rates in kilograms per hectare for oilseed rape

N Index	*Winter oilseed rape*				*Spring oilseed rape*			
	Sandy soils and shallow soils on chalk		*Other soils*		*Sandy soils and shallow soils on chalk*		*Other soils*	
	Seed bed	*Top dressing*	*Seed bed*	*Top dressing*	*Seed bed*	*Top dressing*	*Seed bed*	*Top dressing*
0 straw incorporated	30	160	30*	140*	30	90	120	0
0 straw baled	0	190	0	170*	30	90	120	0
1	30	130	30	60	30	60	70	0
2	0	40	0	40	30	40	0	0

* Increase to a total of 200 kilograms per hectare for early established crops when there is no serious pest damage or weed competition and a yield of more than 3.5 tonnes per hectare is probable.

Top Dressings

Top dressings on winter oilseed rape should be applied in late February or early March.

Top dressings on spring oilseed rape should be applied by early May. Germination can be reduced on droughty soils if all the nitrogen is applied in the seedbed.

Since the nitrogen requirements of crops are greatly affected by the weather during the growing season, which cannot be predicted, it is inevitable that even the most precise recommendations will be high for some years, and low for others.

In three consecutive years the yields shown in Table 2.5 in tonnes per hectare were obtained in an experiment on a chalky soil in continuous barley.

Table 2.5   Seasonal variations in response to nitrogen

Nitrogen rate		Year		
kg/ha	1	2	3	Mean
0	1.44	2.10	1.54	1.69
56	3.49	3.19	3.10	3.26
113	4.37	4.20	3.53	4.03
169	4.41	4.10	3.35	3.95
225	4.74	4.07	2.92	3.91

The field and cropping system were the same for each year, so the variation in yield between years was due to seasonal differences only. Even though the 225 kilogram rate was just profitable in the first year, the overall optimum was only about 100 kilograms, and no more than this amount would pay in the long run. Even when a response occurs to such a high rate in one year, most of the response is realised using the average optimum amount and it is uneconomic to exceed it. The yields were low by current standards but were typical of varieties grown in the seventies when the experiments were done.

Types of Fertiliser

The nitrogen in ammonium fertilisers such as sulphate of ammonia or in injected ammonia (anhydrous or aqueous) does not move in the soil water in the same way as nitrate. So these types cannot be leached from the soil and they are not so quickly available as nitrate. However, these forms of nitrogen are soon converted to nitrate by nitrifying bacteria. This process is rapid in warm soil but quite slow while the soil is cold. Consequently ammonium and ammonia fertilisers are slower acting when applied in the early spring, while the soil is cold, but when the soil is warm there is very little difference. Ammonium nitrate has half its nitrogen in the ammonium form and half in the immediately available nitrate form.

When applied as a top dressing to chalky or very light soils there is a risk of loss of ammonia into the air from sulphate of ammonia and also from urea, which rapidly changes into ammonia. However, there is no risk of loss when these materials are worked into the soil.

PHOSPHORUS

The behaviour of phosphorus in the soil is very different from that of nitrogen but is similar in one respect. Some organic phosphorus is converted into inorganic phosphate each season by the action of soil organisms in a process similar to the formation of nitrate, but crops obtain only a small proportion of their requirements in this way. Unlike nitrate, phosphate does not move in the soil water but becomes insoluble. None is leached from the soil and so unused phosphate fertiliser accumulates. Plants can feed on these accumulated residues and do not have to rely on fresh phosphate from the breakdown of organic matter. Most soils have considerable reserves of inorganic phosphate built up from years of fertiliser application.

Phosphate and Soil Type

The availability to crops of phosphate accumulated from fertiliser residues varies greatly between soils. Phosphate is adsorbed by clay, and until the capacity of the clay to adsorb it approaches saturation over the years the fertiliser phosphate applied is converted into much less available forms. The adsorption of phosphate by the soil is analogous to absorption of water by a sponge. When a sponge has absorbed a small amount of water it is very difficult to extract any water from it, but when the sponge is wet, water can easily be removed from it. In a similar way soils first strongly adsorb phosphate, but as the soil approaches saturation it is held progressively less tightly and becomes more and more easily extractable by plants.

As clay soils can adsorb much more phosphate than light soils, they generally have less readily available to plants than sands or loams which are usually quite rich. However, lime-rich soils like the chalky boulder clay in East Anglia and some Lias clays in the Midlands and the West adsorb less strongly than acid clays and become quite well supplied with available phosphate. In contrast acid soils, particularly the shale soils of Wales and northern England, quickly convert fertiliser phosphate into unavailable forms. Levels of available phosphate do not build up to any extent and they always test low by soil analysis. On such deficient soils each crop has to be generously manured, and it is advantageous to place the fertiliser phosphate near the seed so that the crop can absorb all it requires before the soil converts it to less available forms.

Crop Response

Phosphate deficiency is rare in arable crops these days owing to the accumulation of fertiliser residues over many years. Plants suffering from phosphate deficiency are stunted and lack vigour. The foliage is dull in colour and may have bluish tints. In severe cases the old leaves wither.

Crops vary greatly in their responsiveness to phosphate fertiliser and their ability to extract it from the soil. Cereals give little or no response except on deficient soils and are good extractors, so manuring is mainly a matter of putting back the nutrient removed in the crop. Grass is even less responsive, but potatoes on the other hand give large responses except on very rich soils. Most other crops are intermediate between these extremes. Table 2.6 gives

Table 2.6 Recommended phosphate rates in kilograms per hectare for some arable crops

	Soil/phosphorus analysis index					Kg phosphate per hectare removed in average crop	
	0	1	2	3	over 3		
Early potatoes	350	300	250	250	200	Tubers	20
Maincrop potatoes	350	300	250	200	100*	Tubers	40
Sugar beet	100	75	50	50	0	Roots	30
Carrots	250	200	150	100	50	Roots	30
Kale	100	75	50	25	0	Whole crop	75
Oilseed rape	75	40	40	40	0	Grain	50
Cereals	75	40†	40†	40†	0	Grain	45
						Straw	5

* Reduce to 40 at index 5 and nil at index 6 and above if 7 kg/ha foliar phosphate (11 kg/ha mono-ammonium phosphate) is applied at tuber initiation.
† Increase by 20 kilograms per hectare if yields are consistently above average.

general recommendations in kilograms of phosphate per hectare for some common crops and lists the approximate amount removed in harvested crops. To obtain maximum yield of responsive crops it is necessary to apply much more phosphate than is taken up in the crop. Potatoes often respond to a foliar spray containing water soluble phosphate applied at tuber initiation.

Rotational Manuring
On soils well supplied with phosphate, rotational manuring can be practised. If the soil tests moderate or high by soil analysis, phosphate can be applied once every three or four years without loss of yield. For example, in rotations of demanding crops such as roots and undemanding crops such as cereals the phosphate given to the root crop can be increased by the requirements of the next two or three cereals. Generally, though, normal manuring of potatoes leaves enough residual phosphate for one or two cereal crops. Over the rotation as a whole the amount of phosphate applied should at least equal the amount removed in crops.

Depletion of Available Phosphate
If no phosphate is applied to a soil testing high or moderately high in available phosphate (analysis index 3 or 4) no reduction in crop yield can be expected for several years except in the case of potatoes which are particularly responsive to phosphate. In an experiment on a silt soil, yield of sugar beet was unaffected by omission of phosphate fertiliser until the eighth year and wheat was not affected until the eleventh year. Run-down of sandy soils would be somewhat faster and the depletion of clays rather slower.

Types of Fertiliser
The types of phosphate in most fertilisers are completely water soluble. The commonest straight fertiliser is triple superphosphate, which is also the main phosphate component of compounds. The other main phosphatic component of compounds is ammonium phosphate which is water soluble like triple supers. Although water soluble phosphates are readily available to crops they also rapidly react with phosphate-deficient soils to become unavailable. Some compound fertilisers manufactured in other countries contain nitrophosphates which are only partly water soluble. Nitrophosphates are as effective as the water-soluble forms for all crops except potatoes.

The better types of ground mineral phosphate (North African

and Israeli) are suitable for grass, swedes, turnips, kale and rape on acid phosphate-deficient soils but not on alkaline or recently limed ones. Calcined (heat-treated) rock phosphates can be used for grassland on well-limed as well as acid soils.

Because it is an organic source of phosphate, bone meal has often been thought to be preferable to so-called 'artificial' fertiliser. In fact it is a very insoluble and unavailable form and is not very effective.

POTASSIUM

Potassium, the third major nutrient, resembles phosphorus in being adsorbed by the soil clay. It is less strongly held, however, and can be leached from very sandy soils if they are over-generously manured. In contrast with phosphorus it is usually deficient in sandy soils and often abundant in clays since it is a constituent of clay minerals.

Crops absorb a large amount of potassium during the early stages of growth and on rich soils this can be considerably more than the crop's requirement. Such 'luxury uptake' is wasteful when the crop is harvested green as in silage, kale or vegetable crops. However, in cereal crops or root crops like potatoes or sugar beet most of the potassium returns to the soil with the leaves or by leaching from the leaves. In effect potassium is 'lent' by the soil and after repayment it is ready for loan to the next one.

Potatoes are highly responsive to potash fertiliser while cereals are only slightly responsive unless the land is deficient. Most other crops are intermediate so the order of responsiveness is much the same as for phosphate.

Symptoms of Deficiency

After years of fertiliser application most arable soils are quite well supplied with potash, although light soils do not retain a large reserve and can be quickly run down. Symptoms of deficiency are uncommon and when seen they are usually on light land which has had a history of inadequate manuring or has been depleted by carting off a grass or fodder crop. Deficient plants usually have the edges or tips of the older leaves scorched or withered. The scorching is usually brown in colour, but in barley the tips of the older leaves are white in colour resembling frost damage. In potatoes the haulm turns a dull bluish green, and as the deficiency progresses the haulm becomes bronzed in colour with scorching of the older leaves round the edges and between the veins. In severe cases the haulm collapses completely.

Application of Potash

Table 2.7 gives general recommendations in kilograms of potash per hectare to give optimum yield of some common crops together with the approximate amount removed in harvested crop.

Table 2.7 Recommended potash rates in kilograms per hectare for some arable crops

Crop	Soil potassium analysis index					Kg potash per hectare removed in average crop	
	0	1	2	3	over 3		
Early potatoes	250	200	150	75	75	Tubers	90
Maincrop potatoes	350	300	250	150	100	Tubers	185
Sugar beet	200	100	75	75	75	Roots	80
Carrots	150	100	75	0	0	Roots	100
Kale	100	75	50	50	50	Whole crop	250
Oilseed rape	75	40	40	0	0	Grain	30
Cereals (straw ploughed)	75	40	40	0	0	Grain	35
Cereals (straw removed)	75	40*	40*	0	0	Straw	35

* Increase by 20 kilograms per hectare if yields are consistently above average.

While it is important to supply adequate potash, over-generous applications can be harmful; for example, germinating seeds may be damaged unless the fertiliser is thoroughly incorporated into the soil. Excessive potash can accentuate a deficiency of magnesium and also to some extent boron deficiency.

Of the two fertilisers available—muriate and sulphate—the former is preferable for most crops because it is cheaper and just as effective. The sulphate gives potatoes a slightly higher dry-matter content and so is desirable for potatoes grown for manufacture of chips. Some fruit crops such as redcurrants are very sensitive to chloride (muriate) so the sulphate is advisable when high rates are necessary.

Rotational Manuring

On soils well supplied with potassium, undemanding crops like cereals will not give any response to potash fertiliser, so manuring is just a matter of putting back what is taken off in the crop. On such soils rotational manuring can be practised for rotations of root crops and cereals. The potash for the two or three following cereal crops can be applied to the roots in the same way as for phosphate. However, if one crop in the rotation is a green fodder crop or grass cut for hay or silage, it should be the last in the sequence. If it were given potash for the succeeding crops, this would be wasted due to luxury uptake. Rotational manuring is not suitable on deficient soils, because the potash applied becomes strongly adsorbed and not readily available. On such soils placement of fertiliser near the seed is advantageous.

Depletion of Available Potash

If no potash is applied the rate of depletion of available potash will depend on soil type and crop removal. Reserves are low on sandy soils and these can be reduced to deficiency level by a single green forage or hay crop. When only grain is removed depletion is slow and even sandy soils will take about five years to reach deficiency level. Depletion of clays is very much slower.

On naturally potassium-rich soils like some of the chalky boulder clays and Lias clays, potash is not required for undemanding crops like cereals. Natural weathering of the clay minerals results in rapid replenishment of the supply of potassium available to plants. However, potash fertiliser is generally necessary for demanding crops like potatoes even on rich soils.

MAGNESIUM

Although a major nutrient, magnesium is required by crops in considerably smaller amounts than nitrogen or potassium and it does not become strongly adsorbed in the soil like phosphorus. Consequently, it is not necessary to apply magnesium to every crop and the majority of the soils have reserves large enough to make it wholly unnecessary. Magnesium is slowly lost from the soil in drainage water as well as in crops, so soils without large reserves gradually become deficient. This depletion has accelerated in recent years because the supply of farmyard manure which contains a lot of magnesium has declined sharply in arable areas. Many of the light soils in eastern parts of the country are deficient and it has become necessary to apply magnesium fertilisers to some crops.

Sensitivity to magnesium deficiency varies greatly between crops. Root crops, particularly sugar beet, but also carrots and potatoes (colour photos 6 & 7), are quite susceptible as are many horticultural crops. On deficient soils magnesium must be applied regularly for such crops. Cereals are not very susceptible, but responses have been recorded on very deficient soils. In rotations of roots and cereals grown on deficient soils magnesium should be applied to the roots; residues left will be sufficient for the cereals.

Deficiency Symptoms

The symptoms of magnesium deficiency are very characteristic and consist of yellowing between the veins mainly on the older leaves. The yellow areas turn brown when the deficiency is severe. The large leaves of crops like sugar beet or kale develop a 'marbled' appearance.

The occurrence of magnesium deficiency symptoms can be transitory and it does not always follow that application of magnesium is necessary. Damage to roots may induce symptoms, and yellowing due to temporary magnesium deficiency is quite common in cereals suffering from adverse soil conditions. This occurs during the early stages of growth before the surface feeding roots have properly developed.

Antagonism of Potassium and Calcium to Magnesium

The amount of available potassium in the soil affects the ability of plants to extract magnesium. Excessive amounts reduce its uptake, and it is possible to induce or aggravate magnesium deficiency by the liberal use of potash. The ratio of available potassium to available magnesium is important as well as the actual level of magnesium. In agricultural crops the risk of magnesium deficiency is increased if this ratio is greater than about 5. For horticultural crops this ratio should not be much greater than 2.

The potassium/magnesium ratio is also important to grazing animals. For dairy cows not given supplementary magnesium there is a risk of hypomagnesaemia (grass staggers) unless this ratio is considerably less than 1. On lime-rich soils calcium is also antagonistic to magnesium, so it is difficult to build up the soil magnesium to a high enough level for safety.

To minimise the risk of staggers it is important to avoid excessive application of potash to grassland. Potash should never be applied in the spring when the risk is at its greatest, since luxury uptake can seriously increase the chances of trouble.

Application of Magnesium

On land which needs periodic liming magnesium is often most

cheaply supplied by using magnesian lime in place of some or all of the ordinary lime. When lime is not needed, fertiliser such as Kieserite or calcined magnesite can be used. Table 2.8 gives recommendations in kilograms of magnesium per hectare for some common crops and the approximate amount removed in harvested crop.

Table 2.8 Recommended rates of magnesium in kilograms per hectare for some arable crops

| Crop | Soil magnesium analysis index | | | Kg magnesium per hectare removed in average crop | |
	0	1	over 1		
Early potatoes	100	50	0	Tubers	2
Maincrop potatoes	100	50	0	Tubers	7
Sugar beet	100	50	0	Roots	7
Carrots	60	30	0	Roots	4
Kale	80	40	0	Whole crop	25
Cereals	60*	0	0	Grain	6
				Straw	2

* Applied once every three years in continuous cereals. Magnesium applied to a more responsive crop in rotations.

SULPHUR

Until recently there was never any need to apply sulphur for crop nutrition in Britain because adequate amounts were deposited in rain, which contains sulphuric acid formed from the burning of sulphur-containing fossil fuels. As the prevailing winds are from the west and from the sea, there tends to be less sulphur in rain in western Britain than in the east. With the recent gradual reduction in sulphur pollution, amounts deposited and retained in soils are not now sufficient for all crops in all areas. Double low varieties of oilseed rape, a crop with a high sulphur requirement, now give yield responses to sulphate fertiliser when grown on light soils in some moderately high rainfall western areas such as Somerset, Dorset, Northumberland and Cumbria and in eastern Scotland north of the industrial belt. Responses are less likely to be obtained in low rainfall eastern areas because less sulphate is leached out of the soil and the rain generally has a higher sulphur content. However, sulphur deficiency is starting to show up in oilseed rape in these areas also (colour photo 5). The symptom of sulphur deficiency is paling of both young and old leaves. Cut grass yields are occasionally increased by sulphate fertiliser in eastern Scotland and in western high rainfall areas of England Wales.

The cheapest straight sulphate fertiliser is calcium sulphate or gypsum (16–18 per cent S); 120 kg/ha is the recommended treatment when sulphur is deficient. Sulphur is also present in the magnesium fertilisers Keiserite (23 per cent S) and Epsom Salts (10 per cent S) and in sulphate of potash (18 per cent S). 100 kg/ha of the nitrogen fertiliser sulphate of ammonia supplies 24 kg/ha S and is the cheapest treatment; allowance must be made for the 21 kg/ha N also applied. Sulphur is present in some NPKMg compounds and in some foliar-applied products.

LIME

Lime contains the essential nutrient calcium. Even more important, the amount of lime in a soil controls its degree of acidity or pH. When a soil's base exchange capacity is saturated mainly with calcium it is neutral and has a pH of 7. If there is excess calcium present, the soil is alkaline and the pH is above 7, whereas soils short of calcium are acid and have pHs below 7. As well as affecting pH, lime has some effects on soil structure, as described in Chapter 6.

Losses of Lime

Like magnesium, calcium is leached from the soil in drainage water. The quantity of calcium lost depends to some extent on the amount present but mainly on the amount of water draining through the soil, which in turn is controlled by its permeability and the rainfall. It is also affected by the content of sulphuric acid in rainwater picked up from air polluted by combustion of sulphur-containing fuels. On land near large towns lime is lost more quickly than in rural areas. Away from large towns the average loss of lime is about 0.4 tonne/ha per annum but this can be more than doubled where the air is polluted.

Nitrogenous fertilisers have an acidifying effect on the soil unless they contain lime such as the ammonium nitrate lime mixtures. Sulphate of ammonia has the greatest effect. Bacteria in the soil change it to a mixture of nitric and sulphuric acid and one kilogram removes approximately one kilogram of lime. Ammonium nitrate changes to nitric acid and one kilogram of it reacts with about half a kilogram of lime. Organic manures such as poultry manure have a similar effect unless the diet of the birds producing it is rich in lime.

pH and Trace Elements

The pH of a soil controls the availability of essential trace elements such as manganese and also of aluminium, which is not required by plants but is present in all soils. These elements are least available when the soil is alkaline (pH greater than 7), and as the soil becomes slightly acid—pH between 6 and 7—their availability increases. On mineral soils the ideal pH for most arable crops is about 6.5, and on peaty soils 5.8. The corresponding pHs for grassland are 6.0 and 5.5 to help ensure adequate availability of copper and cobalt.

In strongly acid soils manganese and aluminium become too readily available, and sensitive crops such as barley and sugar beet become partly poisoned by absorbing too much. When the pH falls below 5 in extremely acid soils most crops are adversely affected by absorption of excessive amounts of these elements. The dominant effect of shortage of lime is manganese and aluminium toxicity rather than a deficiency of calcium. Consequently, the application of small amounts of calcium in the form of sprays will not benefit crops suffering from acidity and top-dressed lime has no effect until it reaches the root zone.

Crop Sensitivity to Soil Acidity

Crops vary greatly in their sensitivity to acidity. Table 2.9 gives the approximate pH below which growth is noticeably restricted on mineral soils. On peaty soils the corresponding value will be about 0.5 of a unit lower.

Application of Lime

As lime is only slightly soluble in water it leaches into the soil very slowly. The only quick way to correct acidity in the topsoil is to mix it thoroughly in by cultivation, although top dressing of lime, particularly hydrated lime, gives some benefit to crops suffering from acidity. Hydrated lime is more soluble than ordinary lime and so washes into the soil more quickly. Acidity in the subsoil is also harmful and when it is severe the downward movement of roots can be completely prevented.

Correction of subsoil acidity is only possible by the slow process of leaching from the topsoil, so it is essential to apply maintenance dressings of lime before the topsoil becomes acid enough to reduce

Table 2.9   Critical pH values

Crop	Critical pH	Crop	Critical pH	Crop	Critical pH
Apple		Clover, red	5.8	Potato	4.8
(established)	4.9	Clover, white	5.5	Rape†	5.5
Apple (newly		Cocksfoot	5.6	Raspberry	5.4
planted)	5.5	Kale	6.0	Redcurrant	5.5
Barley	5.8	Lettuce	6.0	Rye	4.8
Bean	5.8	Lucerne	6.1	Ryegrass	5.0
Beet	5.8	Mangolds	5.5	Strawberry	5.0
Blackcurrant	6.0	Mustard	5.3	Swede	5.3
Brussels sprouts	5.6	Oats	5.0	Timothy	5.2
Cabbage	5.3	Onion	5.6	Turnip	5.3
Cauliflower	5.5	Pea	5.8	Wheat	5.3

† See colour photo 1.

crop yield. The amount of lime required to correct the acidity due to a particular pH varies greatly between soils. The lime requirement of a sandy soil at pH 5 will be much smaller, because of its smaller base exchange capacity, than that of a clay at pH 5. Table 2.10 gives the approximate amount of ground limestone or chalk required to correct a range of pH values in different types of soil.

Table 2.10   Lime requirement of ground chalk or limestone tonnes per hectare for 150 mm depth of soil

pH	Sands	Loams	Clays	Peats
6.0	2.5	4	5	0
5.5	5	8	10	6
5.0	7.5	12	15	12
4.5	10	16	20	18
4.0	12.5	20	25	24

If lump chalk is used, it is necessary to increase rates by about one half if the pH is much below 6.0 but liming will be needed less frequently.

Liming Problems

It is a good policy always to lime before the soil becomes acid enough to affect crop growth, because it is very difficult to correct acidity quickly for a sensitive crop. This is because the lime must be thoroughly incorporated in the root zone to allow unrestricted root growth. If a soil has been allowed to become acid, it is advisable to lime, grow a less sensitive crop and allow a year for the lime to work.

Particularly on light land acidity tends to occur in patches, so it is necessary to lime for the most acid parts of the field and this inevitably means that some parts become over-limed. Although undesirable because this can lead to deficiencies of manganese and boron, it is less harmful than acidity to arable crops because the deficiencies can be quite simply rectified.

If acid patches occur in crops, it is sensible to top dress them with lime because the patches are difficult to locate without a crop. Liming the patches helps to even up the pH of the field.

Although it is almost impossible to detect small patches of soil needing lime, unless a crop is suffering from acidity, it is possible to pick out large areas with the aid of a bottle of 'indicator'. The most common indicator is a solution of a mixture of dyes which is blue when the pH of the solution is high and changes to green, yellow and finally red as the pH becomes progressively less alkaline and finally acid (see Appendix 1). It is used by adding a small amount of soil to some indicator in a white porcelain dish, gently agitating, allowing the suspension to clear and then simply comparing the colour with a standard colour chart.

SODIUM

Sodium, a constituent of sodium chloride (common salt), is an essential nutrient to a few species of plants which evolved under saline conditions. The most important in Britain is sugar beet, which needs sodium for maximum yield although it will grow satisfactorily without it. Sodium can substitute for most of the potash requirements of beet and also for part of the potash requirements of a few other crops, e.g. barley, for which it is not an essential nutrient. Sodium is beneficial for carrots, as yields may be increased and quality improved.

There are no clearly recognised symptoms of sodium deficiency. Even when yield is affected by lack of sodium a crop of beet looks perfectly normal. The presence of a large amount of sodium can greatly reduce stability, so over-generous use of salt is unwise on any soil. Its use is risky even in small amounts on difficult unstable soils such as silts and sandy clays, particularly if they are not well drained. The risk on such soils is minimised when they are naturally rich in lime or chalk as the sodium is then soon replaced by calcium.

The main risk from application of salt is of surface capping which can result in poor germination. Also, if it is applied shortly before drilling, germination can be adversely affected by a high concentration of salt in the soil solution. Both risks are minimised by applying it in the autumn.

CHAPTER 3

PLANT NUTRITION
The Minor Nutrients

The minor nutrients, commonly known as trace elements, are so called because they are required by crops in very small amounts. Although minor in amount required, they are not minor in importance. All are as essential to plants as the major nutrients and the effects of deficiencies are often severe. The minor elements known to be essential in plant nutrition are manganese, iron, copper, zinc, molybdenum and boron.

Deficiencies of all these nutrients except zinc are known in this country and the most widespread is manganese.

The minor nutrient content varies widely between soil types. Coarse sandy soils tend to be poorly supplied with trace elements, while most heavy soils have adequate amounts, although there is not always sufficient readily available to crops.

Deficiencies are favoured by droughty conditions. As these are generally most pronounced on sandy soils which also tend to be short of trace elements, most severe deficiencies occur on the sands. The occurrence of minor element deficiencies is greatly affected by the soil's content of lime. Most trace elements become less available to crops as the lime content of the soil increases.

Concern has often been expressed in recent years about depletion of reserves of minor elements by modern farming. Small amounts are removed in every crop and much less is returned to the land in the form of farmyard manure than in earlier mixed farming systems. Compound fertilisers have become more concentrated and contain fewer trace element impurities.

While these statements are undoubtedly true, it is also a fact that most soils contain sufficient of most of the minor elements to supply crops indefinitely. There is no justification for applying expensive trace element mixtures to crops. Such mixtures, in any case, usually do not contain sufficient to correct serious deficiencies. On soils which are not well endowed the occurrence of a trace element deficiency is easily diagnosed and just as easily rectified by applying a spray containing the deficient element. Application of one or more minor elements has become routine farming practice on some naturally deficient soils.

DIAGNOSIS OF MINOR ELEMENT DEFICIENCIES

Minor element deficiencies are commonly diagnosed by the symptoms they cause in crops, which are very characteristic and usually easy to recognise. Analysis of the soil is useful for picking out soils which are deficient in copper and boron. In the case of copper, yield benefits can be obtained when the deficiency is too slight for obvious symptoms, so it is important to know if the soil is deficient. Boron deficiency does not usually show up until quite late in the season when it cannot be treated without crop damage, so again it is necessary to know if the soil is deficient.

Analysis of the soil cannot predict with any degree of certainty the occurrence of a deficiency of manganese or iron, although soils where these deficiencies are a possibility can be distinguished from those where there is no risk at all.

Analysis of the plant can help to diagnose manganese or boron deficiencies, but it is not very helpful in the case of copper and useless in the case of iron. Zinc and molybdenum deficiencies can also be recognised by plant analysis. Although useful, such analyses do not always give clear-cut answers because there is always some element of doubt about the exact level at which a plant becomes deficient.

MANGANESE

The total content of manganese varies greatly between types of soil. Most clays and loams contain a lot of manganese, but many sandy soils are not well supplied. This is partly because sand particles generally contain little or no manganese and partly because some of it has in the past been leached out by rainfall under acid forest or heathland conditions. Although the total amount present in the soil affects availability to plants, availability is also greatly affected by the soil's content of lime and organic matter and by its drainage status.

The availability of manganese to plants decreases as the soil pH is increased by liming. Deficiency in crops only occurs on soils which have been over-limed or are naturally rich in lime. Manganese deficiency is also favoured by a high content of organic matter and it is very prevalent on over-limed or lime-rich peaty soils. It is common on poorly drained soils partly because they tend to have high organic matter contents. It is also common on over-limed sandy soils, particularly heathland sands which are dark in colour. Although the deficiency on naturally acid soils can be prevented by not over-liming, it is often impossible to avoid creating over-limed patches in areas of patchy acidity. As any deficiency that occurs can be readily corrected by spraying the affected crop with manganese sulphate, it is better to be generous with lime rather than risk crop failures due to acidity, which cannot be completely remedied in a growing crop. Ideally a pH of 6.5 is required on mineral arable soils and 5.8 on peats. These levels are high enough to avoid loss of crop due to acidity and low enough to make manganese deficiency unlikely.

As well as being prevalent under droughty conditions, manganese deficiency is also encouraged by poor light. In dull springs trouble in affected areas is usually widespread. Soil conditions also have an effect; a loose fluffy tilth makes deficiency more likely (colour photo 3). Healthy plants contain more than 30 parts per million of manganese in the dried leaves. Below this level deficiency is possible, and it is likely below 20.

Correction of the deficiency is best accomplished by spraying the crop with 9 kg per ha of manganese sulphate when deficiency symptoms occur. Up to three sprays at intervals of about three weeks may be necessary in severe cases. Other forms such as manganese oxide or manganese chelates are either less effective or much more expensive than the sulphate. The symptoms vary greatly between crops, but there is usually yellowing between the veins of the leaves and in most crops the plant becomes limp and appears wilted. Crops are usually affected in patches which look pale and floppy. Cereals exhibit lesions near the centre of the older leaves, which tend to go along the vein but cross the veins when the deficiency is severe. In wheat the lesions are white, in oats they are a light rusty colour, often known as 'Grey Speck' (colour photo 2), and in barley they are dark brown. As well as being pale and limp, manganese-deficient potato plants develop small black spots in a regular pattern between the veins of their leaves.

The symptoms in beet are rather unusual in that, instead of becoming limp, affected plants assume a more upright habit. The leaves curl slightly to form a rather triangular shape and the leaves exhibit small yellow spots which are known as 'speckled yellows'.

Manganese deficiency is the cause of 'marsh spot'—severe browning–in the centre of dried peas. Usually one spray when the crop is flowering is enough for control, but in severe cases two or even three sprays are necessary. Applications to the soil are usually completely ineffective for control in any crop unless the manganese sulphate is mixed with superphosphate close to the seed. Control can be obtained by this method but it is much more expensive than spraying since the amounts required are greater.

On slightly acid soils the availability of manganese is normally about right for all crops, but if the acidity is allowed to become severe manganese becomes excessively available and toxicity results. Manganese toxicity is one of the complex effects of soil acidity.

IRON

Most soils contain very large amounts of iron, more than enough to supply the needs of crops indefinitely. The brown colours in well-drained soils and the rust and grey colours in poorly drained soils are all due to oxides of iron. The only soils containing very little within rooting depth are the chalks, and this is one reason why iron deficiency in agricultural crops is only encountered in chalk soils. The other reason is that, like manganese, the availability to crops is greatest under acid conditions and least in the presence of excess lime.

Iron deficiency is occasionally encountered in sugar beet grown on chalk soils, usually chalk marls, when growth is restricted by soil compaction. It is invariably transitory and the symptoms, yellowing of the young leaves, disappear after a week or so. The deficiency can be rectified by application of iron compounds but this does not improve growth, presumably because the soil compaction affects growth more than the deficiency.

Perennial fruit crops are much more susceptible to iron deficiency than arable crops and most of them are unsuccessful on chalk soils. In mild cases it causes intervenal yellowing of the youngest leaves. As it increases in severity the young leaves become almost white and more mature leaves are affected so the tree appears strikingly yellow. Growth is restricted and in very severe cases the branches die back from the tips.

The most susceptible crops, for example pears, are liable to the deficiency on lime-rich clays like the chalky boulder clay as well as on chalks. Usually, though, it does not occur unless there is also some physical defect affecting root growth, as in the case of sugar beet.

Like manganese deficiency, iron deficiency cannot be corrected by applying simple iron salts to the soil because the iron rapidly becomes insoluble and unavailable. However, it can be controlled using 'complexed' or 'chelated' iron compounds which remain soluble in the soil. Foliar sprays of these iron compounds are also effective. The two materials available are known as iron EDTA and Sequestrene 138 iron. The latter material is most effective for soil applications and the amount needed varies from 30 to 110 grams per tree depending on size. Soft fruit require about 15 g/m of row. Treatments last for about three years.

Both materials are equally effective for foliar applications. About four sprays per year are required at fortnightly intervals of 110 litres/ha containing about 230 grams of either material with a wetter.

COPPER

Although copper is an essential nutrient for all crops, deficiency is only known in a few of them on a small number of soil types. The deficient soils are light peats, mainly in East Anglia, heathland sands, e.g. Breckland of Norfolk, Bagshot sands in Dorset and thin humus-rich chalk soils in southern England. Cereals are the main susceptible crop, but deficiency is also known in onions and lettuce. In East Anglia the deficiency in cereals is exhibited by poorly filled ears and by the bleaching and spiralling of the leaf tips and barley awns (photo 3.1) and, when severe, the crop is seriously stunted. On the chalk soils of southern England the symptoms are very different. Wheat crops become a dark olive green, eventually becoming almost black, and produce shrivelled grain or empty ears. In barley the discoloration does not occur but ears are partly or completely empty. The straw below the ear can be weak and result in premature necking over.

Copper deficiency is favoured by high levels of organic matter and lime and it resembles manganese deficiency to some extent. However, it differs in that it only occurs on soils containing very small amounts of copper and cannot be induced by high levels of lime and organic matter on soils well supplied with copper. It also differs in that on the light peats it can occur on strongly acid soils.

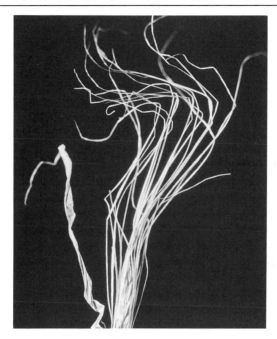

Plate 3.1   Copper deficiency in
barley   (Crown copyright)

Remedying the acidity may partially correct copper deficiency, but over-liming can make it more serious. Correction of copper deficiency may be accomplished by applying copper sulphate to the soil at rates of about 60 kg/ha or just as effectively by applying 2 kg/ha of the fungicide copper oxychloride as a foliar spray when the crop is well tillered. Soil applications last for several years, but foliar sprays have little effect on the level of copper in the soil and so must be repeated for each crop.

Copper deficiency can occur in cattle and sheep grazing deficient land, but this disorder is most commonly caused by nutritional imbalance due to excessive molybdenum taken up from soil rich in that element.

BORON

As with other minor elements, all crops need boron, but their ability to extract it from the soil varies greatly. Potatoes and cereals, for example, readily absorb all they require from British soils but beet crops and brassicas, particularly cauliflowers and swedes, often suffer from deficiency. Sandy soils tend to be short of boron while heavy soils usually contain sufficient. The occurrence of deficiency is favoured by over-liming and it is not known on acid soils in Britain. It is also favoured by droughty conditions and so is most commonly encountered on light sandy soils in dry years.

The deficiency usually causes a weakness or death of the growing point. In sugar beet, for example, the growing point dies and the crown of the root rots giving rise to the typical 'crown rot' or 'heart rot' symptoms (colour photo 4). Small secondary buds develop around the edge of the rotted crown. In cauliflowers the curd is poorly developed and may rot.

In the leafy brassica crops such as kale or cabbage the centre of the stem breaks down and eventually rots. In swedes and turnips boron deficiency is known as 'brown heart' and results in a brown water-soaked zone in the centre of the root. The symptoms are usually most pronounced in the largest roots. Healthy plants contain more than twenty parts per million of boron in the dried leaves. Below this level deficiency is likely, and very likely below fifteen.

Boron deficiency is easily controlled by application of 22 kg/ha of borax to the soil or by application of 6 kg/ha in the form of a spray fairly late in the season. It is most commonly prevented by use of a 'boronated' compound fertiliser containing a suitable small percentage of borax.

Excess boron is very toxic to all crops, particularly those like potatoes and cereals which easily obtain all they need even from deficient soils. Application of boronated compounds can severely damage these crops, particularly if they are placed near the seed. The satisfactory content of boron in the soil ranges from about 2–7 kg/ha (20 to 70 kg of borax/ha). Above about 7 kg/ha there is a risk of toxicity, so there is only a small safety margin and great care is needed to apply the correct amount.

The 'pulverised fuel ash' produced by power stations burning powdered coal is often very rich in boron. Crops grown on land covered with ash may suffer severe boron toxicity until the excess boron has been leached out over a period of years by the rain. Cereals and potatoes are very susceptible but sugar beet and lucerne are quite tolerant.

MOLYBDENUM

Although molybdenum deficiency in clovers and other legumes is well known in many countries, it is rare in Britain except in cauliflowers which are particularly sensitive. It also occasionally occurs in other brassica crops, particularly lettuce. In cauliflowers the laminae of the leaves are poorly developed and when the deficiency is severe, little more than the midrib remains, which explains the name 'whiptail' (photo 3.2). The plants become blind so no curd is formed.

Plate 3.2 'Whiptail' caused by molybdenum deficiency in cauliflower. Normal plants in background   (Crown copyright)

Molybdenum differs from all other trace elements in being most available under alkaline conditions and least available when the soil is acid. So the chance of whiptail is minimised by ensuring that the soil is kept well limed.Most cases of whiptail occur when the soil has been allowed to become slightly acid, but the cauliflower crop is so sensitive that the deficiency can appear even on high lime soils. Most growers ensure against the disorder by applying

$\frac{1}{4}$ kg of sodium molybdate per hectare to the seedbed in which the plants are propagated. The extra molybdenum taken up in the early stages is sufficient to prevent whiptail.

Molybdenum gives rise to more problems in stock than in crops. Certain soils, like some of the Lias clays and some reclaimed marshes, are very rich in molybdenum. Grass produced is excessively high in molybdenum which results in nutritional imbalance and induces copper deficiency in grazing cattle.

Table 3.1 Summary table of minor element deficiencies and their control

Element	Symptoms	Treatment
Manganese	Intervenal yellowing or brown spotting mainly on the older leaves. Crop appears pale and limp apart from beet which is more upright. In cereals lesions occur near the centre of the leaves.	Spray with 9 kg/ha of manganese sulphate in 200–1000 litres with a wetter as soon as symptoms appear. Peas should be sprayed as flowering commences.
Iron	Yellowing of youngest leaves.	In tree fruit crops 30–110 g per tree, depending on size, of sequestrene 138 iron; 15 g/m² in soft fruit, or 3–5 sprays of iron EDTA—230 g in 1000 litres/hectare with a wetter.
Copper	Bleaching and spiralling of leaf tips and barley awns. Cereal ears are poorly filled. On chalk soils wheat becomes dark olive green.	Spray with 2 kg/ha copper oxychloride fungicide when crop is well established.
Boron	Death of growing point and/or rotting or brown discoloration of root or stem.	Soil application of 20 kg/ha of borax or foliar application of 6 kg/ha of borax or 3 kg/ha of 'Solubor'.
Molybdenum	Poor development and distortion of leaf laminae in brassicae, particularly cauliflowers.	Seedbed application of $\frac{1}{4}$ kg/ha of sodium molybdate.

ZINC

Most British soils have supplies of zinc which are quite adequate, and zinc deficiency has not been confirmed in arable crops.

However, some light chalky soils contain less zinc than American soils on which zinc deficiency regularly occurs in maize. Like manganese, iron and copper, zinc is least available in the presence of excess lime, so a deficiency of zinc in some crops may become a possibility on light chalks. Zinc deficiency could also be a possibility on the type of heathland sand which gives rise to copper deficiency when it is over-limed.

Although zinc deficiency is not a problem, zinc toxicity caused by application to the land of contaminated sewage sludge from industrial towns has caused trouble in a number of areas. It is particularly serious because zinc does not leach out of the soil and once the land has become severely contaminated there is no way of putting it right.

COBALT

Although cobalt is an essential nutrient for grazing animals and the amount present in herbage is important to the livestock farmer, this element is not essential to plants. In common with manganese, iron, copper and zinc its availability is reduced by liming. Most soils are adequately supplied with cobalt but deficiencies occur on a number of soil types which can be corrected by applications to the soil.

SELENIUM

Selenium is another minor nutrient essential for the health of grazing animals. It is deficient on some soils. Correction of the deficiency in stock is not accomplished by applications to the soil.

CHAPTER 4

SOIL WATER

All plant growth depends upon a supply of water and this need has to be satisfied by the action of roots that extract water from the soil in which the plants grow. In the British Isles rainfall during the cold season of late autumn, winter and early spring is considerably in excess of the moisture loss from the soil by evaporation and transpiration. This net gain of moisture to the soil replaces that which has been extracted during the previous growing season, and the surplus either penetrates deeply into the geological strata to replenish underground storage, or is taken away into rivers by drains, or builds up a water table in the soil, or runs off the surface.

In almost all areas the soil contains its maximum quantity of water at the end of the cold period. Thereafter the rate of moisture loss from the soil to the air usually exceeds the rate of replenishment by rainfall, and a water deficit builds up in the soil. As a result the moisture content within the rooting zone decreases, until in dry periods growth may suffer and finally wilting occurs. Water is of such fundamental importance to crop production that an understanding of its role is invaluable to arable farmers and students of agriculture.

THE WETTING OF SOILS

After the summer period of moisture extraction by plants, the soil starts to wet up from the surface as rainfall exceeds extraction. In soils without deep cracks rain wets the profile progressively from the surface as a wetting front moves into the soil. In clay soils, with deep cracks, wetting of the soil occurs irregularly and in exceptional circumstances water can be lost out of drains while the majority of the profile is still dry. As the soil continues to wet up, a point is eventually reached when it can hold no more, and in free-draining soils any extra water is lost from the profile. This point at which drainage starts is known as field capacity (FC). Contrary to common belief soil is not saturated at FC but contains appreciable quantities of air in the coarser pores, i.e. larger than 0.05 mm (see Table 4.2). Soils at FC feel very moist to the hand. Research on clays has demonstrated that provided the surface is not sealed, water infiltrates faster into direct drilled land than it does into deep cultivated or ploughed land. By inference direct drilled soil profiles reach field capacity earlier than their cultivated equivalents.

WATER AVAILABLE FOR PLANT GROWTH

Water in soil at FC is only loosely held and easily extracted by plant roots, but as more and more water is removed from the soil from progressively smaller pores the point is reached at which the maximum suction that roots can develop balances the energy with which water is held by the soil. At this point water removal falls to negligible quantities, plants lose turgor and wilting occurs; the soil moisture content is known as permanent wilting point (PWP), and at this point soils feel nearly dry or only very slightly moist. The quantity of water held in the soil between the two limits FC and PWP is known as the available water capacity (AWC) and is usually expressed as mm of water per 300 mm of soil (see Table 4.2).

In many well-structured free-draining soils, rain falling on land already at FC will have left the profile after 48 hours, returning the soil to FC again, but in soils with impeded drainage, extra water may take weeks to drain, and for these soils there is no realistic FC. Silty soils tend to have a greater number of intermediate-sized pores than sands and clays, and consequently water drains down the profile for weeks and sometimes months rather than the two-day period mentioned above. Although plants can make use of all water in the available range provided roots have access to it, the strongly retained water is less easily extracted and cannot sustain the same rate of plant growth as water less tightly held. In consequence irrigation practice aims to replenish soil moisture reserves well before the moisture deficits reach PWP.

The AWC of free-draining soils is a very important characteristic, which, when taken in conjunction with average rainfall, gives a good measure of the cropping potential of land. The major factors determining the AWC of a soil are its texture, depth and structure. Average figures for a number of different textures are given below. Predictably sands hold the least and peat the most, but it is less widely realised that soils with high proportions of fine sand and silt-sized particles hold more available water than clay soils.

A. *Soils with low AWC—less than 40 mm per 300 mm depth:*
   All sands and loamy sands (less than 10 per cent clay), except loamy fine sands.
B. *Soils with medium AWC—more than 40 mm but less than 60 mm per 300 mm depth:*
   Loamy fine sands, all sandy loams, sandy clay loams, clay loams and clay.
C. *Soils with high AWC—more than 60 mm per 300 mm depth:*
   Very fine sandy loams, silt loams, peats and any soils with high dyke water tables.

In soils with different textured layers the AWC can be calculated by adding the AWCs for the individual depths.

The effect of soil structure on AWC is less easily defined, but in general where conditions are more compact FC will be lower and the quantity of water available reduced. The absence of root penetration into compacted soil will accentuate the reduced availability of water to plant roots (see page 74). In cases where structure is so poor that water movement is made very slow, this could be interpreted as an increased AWC; however the damage resulting from restricted rooting usually cancels out any advantage from increased water. Moisture content of growing crops varies from 50 to 98 per cent depending on the nature of the plant material, but the total requirement of a crop for water during the growing season is many times greater than these figures suggest. This is because there is a continuous loss of moisture from the leaves of a crop canopy at a rate determined by the meteorological conditions. This process, known as evapo-transpiration, is inevitably associated with growth and as a spin-off the evaporation process helps to keep the crop cooled on sunny days. In temperate climates for each unit of dry matter accumulated by crops, 200–250 units of water is transpired through the crop; thus a 10 tonne/ha grain crop (+ 10 tonnes straw) has used about 4,000 tonne/ha of water, i.e. about 50 cm.

The AWC of a soil gives a useful indication of whether or not water shortage is likely to reduce yields. For example, in eastern England the average maximum summer soil moisture deficit under cereals, potatoes, grass and sugar beet is 125–150 mm of water.

Soils in group A with low water-holding capacity (less than 110 mm/900 mm) will suffer from drought almost every year, whereas soils in the high group (more than 190 mm/900 mm) will suffer from drought much less frequently and only in very dry years. The estimated effect on crop yields is shown in Table 4.1.

With irrigation or in wetter areas of the country yields from low AWC soils will be substantially higher.

Table 4.1 Estimated potential yields in areas of 500–625 mm rainfall

	Soils of low AWC	Soils of high AWC
Sugar beet	5–6 t/ha sugar	12–13 t/ha sugar
Potatoes	25–30 t/ha	80–90 t/ha
Spring barley	4–5 t/ha grain	8–9 t/ha grain
Winter barley	5–6 t/ha grain	9–10 t/ha grain
Winter wheat	4–5 t/ha grain	10–12 t/ha grain
Winter oilseed rape	$3\frac{1}{2}$–$4\frac{1}{2}$ t/ha	$5\frac{1}{2}$–$6\frac{1}{2}$ t/ha

WATER MOVEMENT IN SOILS

When heavy rain falls on a soil already at FC the coarse pores and fissures that are filled with air are temporarily filled with water and the soil becomes saturated. The soil water in excess of FC is known as drainage or gravitational water (see Table 4.2), and in heavier soils is less in quantity than the available water. The rate of movement of drainage water is dependent on the continuity of large pores and fissures through the soil profile, and in particular on the permeability of the slowest layer. In some cases this may be at the surface and can cause flooding, but in others it may be lower in the profile and may result in a build up of water. In well-structured soils the permeability is usually fast enough to prevent even heavy rainfall building up in the soil, but on poorly structured soils, and those that easily run and cap at the surface, this is not the case.

Once drainage water has gone from the soil the remaining

Table 4.2 Classes of soil water

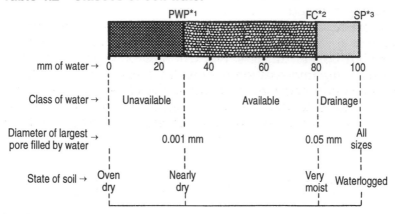

*1 PWP = permanent wilting point
*2 FC = field capacity
*3 SP = saturation point

moisture moves much more slowly. This movement is known as unsaturated or capillary water movement and includes water movement upwards from a water table. The rate of unsaturated water movement is fastest in fine sandy loams, peats and silts and is slowest in sands or clays. In general, this type of moisture movement is so slow that roots have to actually explore the layers of soil holding available moisture before much uptake takes place. However, in cases where there is a water table in the soil below the rooting zone, moisture movement upwards can be appreciable, particularly in fine sands, peats and silts. If the water surface is several feet lower than the depth of rooting, this contribution becomes slight.

LOSS OF WATER FROM THE SURFACE

Water is lost from the soil surface by direct evaporation from wet soil and by transpiration through the leaves of plants. The combined process is known as evapo-transpiration. In general in Britain only 10–15 mm of water is lost from a wet soil by evaporation before the surface becomes so dry that further loss is negligible. This means that land in fallow can only dry out in the top few inches, unless cultivations are used to bring up wet soil from below. However, in land with a water table above 600 mm in the soil, water loss by evaporation will continue for long periods before the top dries out.

Transpiration is the major source of water loss from soil, and is more effective than evaporation in drying out soils to depth. In one recorded period of eight months, loss of water from land covered by a legume was eight times as great as the loss from similar but uncropped land. It follows, therefore, in those cases where the aim is to dry the soil to depth, i.e. for drainage or subsoiling, that a growing crop is much more effective than bare fallow.

MEASUREMENT OF WATER IN SOIL

The moisture content of a soil can be easily measured by recording the loss in weight of soil samples during drying in an oven. By taking samples at intervals during the season changes in soil moisture content may be estimated, but unless the volume of soil involved is also determined by coring on each occasion the quantity of water lost or gained by the soil cannot be calculated. In practice by far the most convenient method for measuring soil moisture changes quantitatively is a technique using the neutron probe. Fast neutrons emitted by the probe of this instrument are moderated by water in the soil, and the greater the water content the greater the number of moderated neutrons produced. The neutron meter's detector counts the moderated neutrons and so provides a direct measure of the quantity of water in the soil. Measurements are made by lowering the instrument's probe into a permanent metal access tube located to depth in the soil. Sequential measurements during the season provide a moisture extraction pattern and therefore an indication of rooting depth.

SOIL TEMPERATURE AND SOIL WATER

The temperature of soil in the spring has an important influence on time of germination and rate of seedling growth. The greater the amount of water in the soil, the slower its temperature will rise in the spring. This effect is partly due to the heat required to raise the temperature of the extra water, but in greater measure it is due to the heat required to evaporate the extra moisture. The chief reason why an undrained clay is colder than one that is well drained is this cooling effect associated with the larger evaporation of soil moisture. The effect is clearly illustrated in an example from F. H. King's book, *Physics of Agriculture*. On a windy day in April when evaporation from the surface was rapid, the air temperature was 12°C. On this occasion the temperature of drained land at 100 mm was 13°C while adjacent undrained land was only 10°C.

WATER BALANCE AND IRRIGATION OF CROPS

Starting with land at field capacity in early spring it is not difficult to construct a reasonably accurate water balance for use in irrigation systems. The rate at which water is lost from land by evapo-transpiration is directly related to the weather conditions, provided there is a full crop cover. It is assumed that moisture loss from bare ground does not exceed 12 mm, and for the period during which the crop is establishing a cover, the transpiration loss is assumed to be half the potential value. Weekly values for potential evapo-transpiration are calculated from standard meteorological data and listed in the *Farmers Weekly* soil moisture guide, or can be obtained from the local ADAS office. The actual soil moisture deficit is equal to the difference between the calculated loss by evapo-transpiration in millimetres, minus the rainfall in millimetres for the same period.

Suppose that it was required to irrigate grass in eastern England, and that potential transpiration data was available every week. (These figures are obtained by correcting the average potential transpiration for deviations from the average recorded sunshine hours.) If sufficient rain fell in March to maintain field capacity, calculation would begin at the beginning of April. Assuming an irrigation plan of 25 mm at a 50 mm soil moisture deficit the balance is given in Table 4.3.

The balance would continue until September when need for irrigation becomes less.

If the field had been in main crop potatoes rather than grass, irrigation would obviously not have been needed until later. If the tubers were planted on April 14, then before the crop emerged the maximum deficit would not exceed 12 mm. After emergence a factor for percentage crop cover is introduced and on average water will be extracted by the crop at 50 per cent of the potential transpiration rate. From about the beginning of June, or when full crop cover is established, the extraction will be at the full potential transpiration rate.

In a similar way the total annual water requirement for irrigation

Table 4.3 Water balance for irrigation of crops

Week ending	Rain (mm)	Transpiration (mm)	Deficit (mm)
April 7	Nil	13	13
April 14	13	19	19
April 21	38	19	Nil
April 28	Nil	23	23
May 5	8	22	37
May 12	Nil	23	60
	(25 irrigation)		
May 19	Nil	22	57
	(25 irrigation)		

on a farm can be calculated. For more detail the reader is referred to
MAFF Technical Bulletin number 16 entitled *Potential Transpiration*.

Irrigation need depends on the crop as well as the soil moisture
deficit, and for very responsive crops the available water content
of the soil is also taken into account. Table 4.4 gives a summary of
recommendations for a number of crops based on data in MAFF
short term leaflet No. 71.

SOIL TYPE AND IRRIGATION

The lower the available water capacity of a soil, the greater the
probability of response to irrigation. In drier areas of Britain light
soils are marginal for production of potatoes and vegetable crops,
but with access to high quality water for irrigation, their potential
can be much improved and the range of crops that can be grown
profitably extended.

The case for irrigation on clay land (40 per cent or more clay) is
usually doubtful, partly because the available water capacity is
medium to high, and also because the risks of crop damage by
waterlogging if heavy rain falls after irrigation, and of damage to
wet land from farm machinery, are substantial. Where irrigation is
used on clays it is worthwhile stopping application while there is
still a 20 mm–30 mm deficit so that rain can be absorbed without
much risk of waterlogging, and also allowing a substantial deficit
to build up towards the end of the crop life, so that the chances
of harvesting under reasonably dry conditions are improved.
Regular irrigation on clays, year after year, should be avoided;

Table 4.4   Irrigation guide for various crops

Crop	Growth period at which to irrigate	Irrigation plan for soils with			Comments
		Less than 40 mm per 300 mm available water	40 mm–65 mm per 300 mm available water	More than 65 mm per 300 mm available water	
Early potatoes	As soon as deficient	25 mm at 25 mm SMD*	25 mm at 25 mm SMD	25 mm at 25 mm SMD	Very responsive crop.
Maincrop potatoes	From the time tubers reach marble size onwards	25 mm at 25 mm SMD	25 mm at 25 mm SMD	25 mm at 25 mm SMD	Relatively shallow rooting and responsive crop. Yield for high AWC soil will be high without irrigation. Controls scab on light soils.
Grass for grazing	Throughout life	25 mm at 25 mm SMD	25 mm at 25 mm SMD	50 mm at 50 mm SMD	Where water is limited, apply 25 mm over as large an area as possible during dry periods.
Sugar beet	When leaves meet in rows until end of August	40 mm at 40 mm SMD	40 mm at 40 mm SMD	No irrigation	A deep rooting crop and only moderately responsive.
Cauliflowers (summer and autumn) and cabbage (summer and autumn)	Throughout life	25 mm at 25 mm SMD	25 mm at 25 mm SMD	50 mm at 50 mm SMD	Where water is limited satisfy deficit about 3 weeks before cutting.
Self blanching celery	Throughout life	25 mm at 25 mm SMD	25 mm at 25 mm SMD	50 mm at 50 mm SMD	Very responsive crop. Nitrogen top dressing can be conveniently applied in irrigation water.
Dwarf French beans	From green bud stage onwards	25 mm at 25 mm SMD	50 mm at 50 mm SMD	50 mm at 80 mm SMD	Avoid raising soil to field capacity immediately before harvest.
Lettuce (summer)	Throughout life	25 mm at 25 mm SMD	25 mm at 25 mm SMD	50 mm at 50 mm SMD	Responsive because of shallow rooting.
Barley, wheat	Late tillering to early stem extension	50 mm at 50 mm SMD	50 mm at 50 mm SMD	No irrigation	Usually only one application will be possible.

* SMD—soil moisture deficit. 25 mm–50 mm of water will sometimes be needed to return soil to field capacity before sowing or planting.

otherwise the absence of strong drying cycles will cause deterioration of structure.

When irrigation is applied to soils with unstable structure, care should be taken to ensure that the rate of application does not exceed the infiltration rate, that the final moisture content is 20 mm–30 mm less than field capacity and that the droplet size is small. In general the application rate to bare soil should not exceed 5 mm per hour or else slaking and surface capping will occur.

QUALITY OF IRRIGATION WATER

Plants grown in saline water are stunted and dark bluish green in colour. Since these symptoms are common to plants growing in saline water irrespective of the actual salt present, it is deduced that the effect is normally caused by the total concentration of salts rather than by the influence of individual salts. Experimental evidence indicates that this 'salt effect' is essentially a water deficiency caused by the osmotic properties of the salt solution. The higher the total concentration of salts in the soil water, the higher the osmotic pressure and the greater the difficulty encountered by roots in absorbing water. At high concentrations of salts, water actually moves out of the roots into the soil.

In addition to the crop damage caused by excess salinity in soil, plants also suffer from the direct scorch due to irrigating foliage with saline water. This effect is particularly noticeable in waters high in chloride.

In Britain saline irrigation water is found in several situations: first, some boreholes are naturally high in salt content; second, in coastal areas sea water contamination of dykes and water tables is common; and lastly river water is frequently used several times over, and this recycling is liable to increase the salinity to danger level. Outside glasshouses, irrigation in Britain is only supplementary to rainfall, so that saline water can often be used without danger to crops, because of the dilution which will take place in the soil. However, it is important to know the sensitivity of the crop to salinity, and the level of salinity of the water, before a safe decision can be reached. Tables 4.5 and 4.6 give information on these two aspects of irrigation. Table 4.6 gives the quantity of saline water that can be safely applied to the different crop groups. Three classes of water quality are specified with low, moderate and high salinity. The quality of irrigation water can only be determined by chemical analysis in the laboratory, where the

Table 4.5 Crop tolerance to saline conditions

Crop tolerance group	Fruit	Vegetables	Field crops	Flowers
Very sensitive	strawberries top fruit gooseberries	French beans peas	clovers	tulips daffodils freesias hydrangeas
Sensitive	redcurrants raspberries	carrots celery lettuce onions	field beans maize cocksfoot	roses gladioli asters
Moderate tolerance	blackcurrants vines	cauliflower broccoli cabbage	potatoes wheat ryegrass lucerne	carnations chrysanthemums
High tolerance	—	asparagus spinach red beet	kale barley sugar beet	—

Taken from USDA Agricultural Handbook Number 60.

electrical conductivity and chloride content are measured, or in the field with the use of a special kit (see Appendix 2).

If the salinity is mainly due to sodium salts, e.g. seawater contamination, then care must be taken not to apply large quantities of water to heavy or unstable structured soils, or structure may be destroyed. Extreme examples of this type of damage are seen when sea water floods over reclaimed sea marshes, with complete loss of topsoil structure in some cases.

Table 4.6   Limit of safe irrigation with saline water

Crop sensitivity*	Salinity of irrigation water†	Safe application for one crop mm of water‡
Very sensitive crops	low	50 mm
	moderate	not suitable
	high	not suitable
Sensitive crops	low	110 mm
	moderate	40 mm
	high	not suitable
Moderately tolerant crops	low	200 mm
	moderate	75 mm
	high	40 mm
Very tolerant crops	low	unlimited
	moderate	130 mm
	high	65 mm

* See groups in previous table.
† Low salinity = conductivity of less than 1,800 μmhos/cm and/or less than 300 mg/1 chloride.
Moderate salinity = conductivity between 1,800 and 3,000 μmhos/cm and/or between 300 and 700 mg/l chloride.
High salinity = conductivity 3,000–5,000 μmhos/cm and/or 700–1200 mg/l chloride.
‡ For application to soil of low–medium water capacity.

CHAPTER 5

LAND DRAINAGE

Over half of the agricultural land in England requires artificial drainage to help remove excess rain falling on the land. On land where natural water movement is obviously too slow a well designed drainage system is an essential part of profitable crop production, and often gives an excellent return on the capital investment. Apart from basin areas like the Fens, where the ground water table is high, the need for drainage is dictated primarily by climate, and it is interesting to consider how this varies over the lowland areas of this country.

INFLUENCE OF RAINFALL

Some time during the autumn or winter period the soil moisture deficit, that has built up during the growing season, is eventually made good by rain and the drains start to run. In wetter areas, for example west Lancashire, the deficit is on average made good by mid-September, but in drier areas (for example, Kent and Essex), this date is delayed until early December. Again in the wetter parts, excess winter rainfall is greater on average, 450 mm in west Lancashire against 125 mm in Essex, and the need for efficient drainage systems in these areas is great every year (see Table 5.1). In Essex and other drier areas, on the other hand, the excess of winter rainfall is so small one year in every four, as to make minimal demands on drainage systems in these years.

At first sight the greater emphasis put on drainage by farmers in the drier eastern part of England, than in the wetter north and west, is a paradox. It can be partly explained, however, by the widespread adoption of arable enterprises in the east on all soil types, whereas in the wetter areas extensive livestock systems have been dominant in the past, and arable crops traditionally reserved for naturally free-draining soils. In recent years intensification of livestock systems, and increase in arable crops in 'grassland areas', have both increased the need for better drainage.

Table 5.1 Excess winter rain in millimetres 1940–59

	Average annual rainfall	Smallest excess	5 years in 20 there are less than	Mean excess	5 years in 20 there are greater than	Largest excess
Morecambe (Lancs)	96	20	16	46	21	83
Bristol	80	13	10	32	19	55
Rothamsted (Herts)	70	6	7	22	11	41
Cambridge	55	2	3	12	6	27
Felixstowe (Suffolk)	53	0	2	10	2	25

Taken from MAFF Bulletin Number 24.

REASONS FOR DRAINING

The importance of removing all excess rainfall at whatever time of year it falls varies a great deal with land use, which in turn affects the economics of drainage in different situations. In Chapter 1 we noted that when soil is saturated with water, plant roots can suffer through poor aeration and from toxic materials which accumulate in these situations. In practice damage of this type only occurs if the soil temperature is high enough for the roots, bacteria and other micro-organisms to be active and using oxygen in the soil. If the soil temperature is below about 5°C, then damage will be negligible. It is for this reason that winter cereals and grass withstand long periods of near saturation in soil during cold winters without significant damage. However, if even a short period of saturation occurs during periods of higher temperatures, the same crops can be severely damaged. This is well demonstrated by top fruit in wet springs, when trees can be killed if the temperature rises and roots become active. Germination is another stage of development more susceptible to waterlogging. Short periods of saturation immediately after sowing can substantially reduce emergence of crops.

Although the direct effect of saturation on root systems is important, there are other more significant reasons for draining.

Capacity of soils to carry machinery falls steadily as the soil moisture content increases, and at moisture levels above field capacity soils can bear only very light weights without rutting. Fields with poor drainage are therefore much more prone to soil damage, from both farm machinery and poaching, and when smearing and compaction occur water movement is slowed down even more. The extra water in the soil results in the land drying out later in the spring, and because extra heat is needed to remove the water the land lies colder for longer (see page 51).

More nitrogen fertiliser is needed to prevent deficiency than on well drained soils, and in the autumn the risk of wet soil at cereal harvest and root lifting is greater, and again the land sustains extra punishment.

Consequently, in poorly drained land we have the right ingredients for a vicious spiral of deterioration in soil fertility and yields. Wet conditions encourage late drilling and damage to soil structure. These cause patchy crops with unhealthy root systems, which are in turn more susceptible to competition from weeds, pests and disease. Poorer crops take less water from the soil, so structure is not improved by strong drying cycles, and the wetter conditions precipitate even more damage from the machinery. This spiral decline in fertility is not just theoretical but is in evidence whenever land drainage has been neglected.

Where land in this condition is taken over, opening ditches and drainage are the crucial factors for reversing the spiral into one of increasing fertility. Judicious subsoiling, weed control and cultivations can then begin to improve crop yields, and in three to four years, with the help of one or two dry autumns, the restoration is achieved.

A similar spiral takes place in grassland farming, but in this case poor drainage results in a degeneration of the species in the sward, with gradual elimination of sown species as they succumb to less productive weed grasses, sedges and rushes, which are better able to survive poaching and wetter conditions. Again, drainage followed either by fertilisation and careful grazing, or by destruction of the old sward and reseeding, will achieve the original level of production in due course.

INDICATORS OF DRAINAGE NEED

Wet areas in fields where tractors bog down, patchy nitrogen-deficient crops giving low yields, excessive growth of weed grasses and greater damage from pest or disease are all indicative of poor drainage in arable land. On grassland, reduced grazing season, worse poaching and replacement of sown species by weed grasses and rushes are usual symptoms of poor drainage.

Examination of carefully placed soil pits, dug to a depth of at least 1 metre, give additional evidence, which is very useful in deciding the cause of the problem and the right solution. Poorly drained soils have duller colours in the topsoil with greyish tones, and higher levels of humus. If the drainage is very poor, the subsoil colour will be predominantly grey throughout, but in land with an alternation of waterlogging in wet seasons and aeration in drier periods the grey colours at depth give way to ochreous and grey mottled soil above, often with black specks of manganese dioxide or 'crowstones'. Free soil drainage is indicated by uniform warm brown or yellow shades throughout the soil. If water is only held up by one particular layer, symptoms of poor drainage will show within and above the layer, while colours indicating free drainage will be found below.

Unfortunately colour as an indicator of drainage status is often permanent and remains largely unchanged long after an efficient drainage system has been installed. However, confusion can usually be resolved by examination of the soil during wet periods in the winter, when water in the profile and soil structure are useful indicators. Outside basin areas with ground water and flush sites, poor drainage is associated with inadequate subsoil fissuring. Careful examination of spadefuls of soil taken from the side of the pit will show if the structures are large, compact and tightly packed with smooth polished faces, indicating slow water movement, or if they are porous and fit loosely together indicating better permeability.

SITUATIONS REQUIRING DRAINAGE AND TYPE OF SYSTEM REQUIRED

There are four different conditions which give rise to waterlogging in soils and it is important to identify which condition applies in a particular situation so that the correct treatment can be applied.

Lowland Basins
In the water table situation, regional water levels must be kept low by a ditch network, sometimes supplemented by pumping or sluices, which control ditch levels. Permeable soils in these situations need

some pipe drainage to pin down the water table, and the spacing will depend on the permeability of the subsoil. When a pit is dug below the ditch water level, these soils rapidly fill with water (fig. 5.1). This type of situation is found for example in the Somerset Moors and Levels, the Waveney Marshes in Norfolk, the East Anglian Fens, Romney Marsh in Kent and in some of the Vale of York Sands.

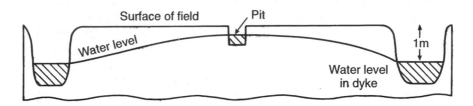

Figure 5.1   Lowland basins. Relationship between water level in dykes and field of moderately permeable soil

If the regional water table is inadequately controlled and the land regularly floods, obviously field drainage should not be contemplated, but where the water table control is satisfactory, the correct design is determined by:

1. The maintained water level in the ditches.
2. The permeability of the subsoil.
3. The variations in land levels.
4. The field size.
5. The crops to be grown.

In practice on level land, if ditch levels are maintained at 1.5 m or below, if soil permeability is high, as, for example, in most deep peat, and if field sizes are small, no pipe drains are needed. But where permeability of the soil is only moderate, drains from 20 to 60 m (2–3 chains) apart are necessary, for most arable crops. The pipes should be placed as deep as the ditches will allow and permeable fill will usually not be needed.

Clayland Areas
The permeability of clay soils when they are fully swollen by winter rains is usually too slow to give efficient natural drainage, and water builds up on the impermeable layers below. Soils formed on Oxford, London and Carboniferous Clays, and also on

many heavy boulder clays, are examples of this type.

It is important to realise that the rate of water movement in subsoils of these clays is often so slow that drains would have to be at 5 m intervals or closer to achieve adequate water control. An intensive system of this type is obviously not economic in modern agriculture, and surface drainage of the stetch, or rig and furrow type, is no longer practicable.

The methods used to deal with this situation involve mole drainage or deep subsoiling, together with pipe drainage. The subsoil treatments provide permeability and intensity, and the pipe drains receive water from the permeability treatment. In this way the pipe lines can be spaced widely at 20–200 m apart, depending on the soil type, surface gradient and type of secondary treatment. An essential part of this system is provision of permeable backfill over the drains to act as a connector between the permeability treatment and the pipes (fig. 5.2 and photo 5.1). The length of life and quality of the permeability treatment is obviously a vital feature of this type of system, and is considered in a later section.

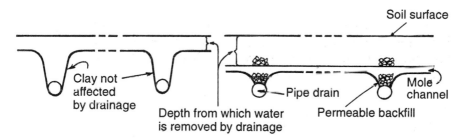

Figure 5.2   Clayland areas. Diagram to show the importance of permeable backfill and moling over drains in clays of very slow permeability

Some clays are particularly well suited for mole drainage, an example being East Anglian chalk boulder clay, which will normally hold a channel for over ten years. In this type of clay, spacing of main drains can be wider, with little risk of the channel 'blowing', provided there is some slope. Other clays, particularly the non-calcareous types, for example London and Carboniferous Clays, and those with an appreciable sand or silt content, e.g. Weald Clay, are less favourable for moling, and channels have a smaller life span. In these soils tile drains need to be closer to avoid the risk of wet areas.

Plate 5.1   The need for permeable fill over drains and a subsoil treatment is demonstrated in this aerial photograph of a drainage trial on a London Clay soil. Winter wheat in spring is best on the mole drained plot. The 'pipe drains only' plot with no permeable fill or moling is no better than the undrained plot   (Crown copyright)

Spring Sites

Very commonly local flush or spring sites occur in undulating landscapes and valley slopes, usually where permeable strata such as sands and gravels, or limestones, overlie impermeable clays (see fig. 5.3). Conversely, water moving laterally over impermeable clays can cause waterlogging at the junction with permeable soils lower down the slope. In both situations the quantity of water to be disposed of may be very large, and precise location and depth of the drains is vitally important to intercept the water moving laterally, at the junction of the permeable and impermeable strata.

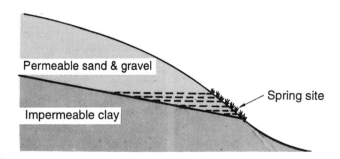

Figure 5.3   Spring sites arise on slopes at junction of permeable and impermeable geological materials

Pans in Soil

Pans of low permeability, either of natural origin or produced by machinery, are frequently the source of slow drainage, causing ponding of water in arable soils (colour photo 18). Obviously it is important to be able to diagnose this type of problem so as not to confuse it with conditions requiring pipe drainage.

In some classes of soil, clay from the upper layers has gradually moved downwards over hundreds of years to form layers or pans of enriched clay content in the subsoil. A good example of this is found in many soils on Keuper Marl, in which periodic deep subsoiling to open up fissures in the heavy layers provides all that is necessary to give satisfactory drainage, provided the lower subsoil is permeable. In other soils thick iron/humus pans have formed in the upper subsoil, preventing root penetration and slowing water movement. Deep busting with a very strong implement will disrupt the pans, but sometimes pipe drainage is needed in addition. Examples of these soils are common in Essex and Herts, where shallow sands and gravels overlie London Clay.

DRAINAGE OF UNSTABLE STRUCTURED SOILS

Some of our most productive arable soils have weak unstable structure, which readily breaks down even during short periods of waterlogging. The brickearths of Sussex, Kent, and East Anglia, the Vale of York Sands, the silt soils of the Wash area and Romney Marsh, and the sandy boulder clays of Suffolk and south Norfolk are examples of this type. Many of these soils are intensively farmed, with few if any livestock, and levels of soil organic matter are frequently low. The weak structure of these soils encourages frequent overcompaction by machinery, and wherever natural subsoil drainage is slow, efficient control of drainage water is essential if farmers are to reduce the risk of soil damage and crop loss in wet years.

Subsoils of undrained marshlands are unripe; that is, they are soft and unconsolidated, containing large quantities of water. Pipe drains remain dry or block with slurried soil soon after installation—often before any water leaves the drain. The only way to 'de-water' these soils is to establish a vigorous crop which will transpire the excess water through its leaves. Shallow ditches every 10–20 m help to give enough drainage for the crop to establish. By extracting water from the subsoil the soil is ripened and stable structure gradually develops. After 2–3 years when this process is sufficiently advanced, pipes can be installed and permanent field drains established. Occasionally in these situations, in what are known as acid sulphate soils, iron pyrites (ferrous sulphide) are found in the subsoil. During drainage these revert to free sulphuric acid, which produces severe acidity in the subsoil, and iron ochre (ferric hydroxide), which blocks drains. Where the initial amount of pyrites is high, reclamation is normally uneconomic.

OVER-DRAINAGE

As discussed on page 61 water tables in lowland basin areas must be kept low with a ditch network supplemented by pumps or sluices. However, during summer the presence of a water table at about 1 m depth is an ideal insurance against drought for all but shallow-rooting crops. Frequently in areas such as the Fens satisfactory drainage of the lowest areas entails 'over-drainage' of the remainder. Partial remedy for these situations can be achieved by isolating areas of land from the main drains with sluices, thereby maintaining higher water tables in the isolated areas.

DRAINAGE AND FARMING SYSTEM

Naturally poorly drained land may need different standards of drainage control, depending on the type of crops grown. The factors involved include the value of the crop, susceptibility of the species to excess water, cultivation requirements for the crop, and time and type of harvesting. Drainage requirements for a number of agricultural and horticultural crops are summarised in Table 5.2. The important point arising from the last column is that different farming systems often need different drainage controls; for example, the same land which for a Cox orchard would need tile drains 10 m apart may only require a skeleton system for cereals, and none at all for beef-raising at low intensity.

OPTIMUM CONDITIONS FOR PIPE DRAINAGE

Experience has shown that if drains are laid in wet conditions with slurried materials in the trenches, efficiency of the system will be much reduced. Wherever possible, these conditions should be avoided by draining earlier in the year, but if they are encountered operations should be postponed until conditions are drier. In addition to the risk of a system being ineffective, drainage in wet conditions invariably damages topsoil structure, and is likely to depress yields in the next crop. There is a very strong case for draining as early as possible after peas, cereals or grass.

MOLE DRAINAGE AND SUBSOILING

The mole plough draws a channel of 50 mm diameter at a depth of 500–700 mm, and in the right conditions fissures the profile above this depth. It is an ideal tool for improving the permeability of heavy clay soils, and for moving water into widely spaced pipe drains. Moles should be drawn at 2–3 m apart depending on the permeability of the soil, and as deep as necessary to ensure the channel is in subsoil of high clay content. For good moling the runs should have an even gradient—not too steep or the channel will erode, not too flat or the channel will block. Beam ploughs are preferable because they even out surface irregularities and give a smooth gradient to the channel. If pulled in good conditions, moles can remain effective for 5–12 years before fresh ones need to be drawn.

Moling is a relatively cheap operation compared with installing a full pipe drainage system, but for satisfactory results careful attention must be given to a number of points. If the subsoil contains pockets of sandier material the length of life will be less, and pipe mains should be closed up to reduce the risk of wet patches forming.

The subsoil at moling depth should be moist enough to form a good clean channel, but not wet or collapse may occur. Ideally moling should be done through a crop when the soil is drying out in May or June; successful moles are pulled later in the year as well, but if heavy rain falls soon after there is a risk of early failure. The soil above moling depth should be dry enough to lift and fissure—this is an essential feature of the technique on very heavy impermeable clays. A system of pipe mains, covered to moling depth with permeable fill, is essential to provide an outfall for the moles, and should be sited at the bottom of slopes and depressions. Optimum spacing of main drains depends on type of soil and gradient of the land; in good moling clays with some slope, up to 200 m can be satisfactory, but in patchy clays with nearly level land, drains may have to be as close as 20–40 m. Ditch outfalls must be deep enough to prevent water from backing up moling channels.

Table 5.2 Drainage requirements of various crops

Crop	Value of Crop (very high–low)	Susceptibility of Crop to Poor Drainage (high, medium, low)	Requirement for Cultivations (high, medium, low)	Requirements for Harvesting (high, medium, low)	Overall Drainage Requirement (very high–low)
Winter cereals and oilseed rape	medium	medium—although they can compensate well these crops are at risk during the wettest period of year.	low—cultivations made during normally favourable soil moistures, and tilth requirements not exacting.	low—harvested when soil moisture content usually not high.	medium
Spring cereals	low	low—growing during drier parts of the year.	medium—seedbed preparations difficult in many seasons and helped by drainage.	low—harvested when soil moisture contents usually not high.	medium
Extensively grazed grassland	low	low—high production not required.	low	low—grazing intensity low.	low
Intensively grazed grassland	medium	medium—high productivity needed, therefore better species must be encouraged, at risk during wettest period of year.	low	high—risk of poaching in spring and autumn is high.	medium
Potatoes (maincrop lifted mechanically)	high	high—even short periods of water-logging in ridges will kill crop.	high—deep tilth needed. Large clods undesirable.	high—lifted during season when drains may be running. Risk of soil damage increases for later dates.	high
Sugar beet	high	medium—will survive short period of flooding.	high—fairly fine even seedbed required to encourage rapid emergence.	high—risk of soil damage high in wet autumn.	high
Field beans	low	medium—disease damage encouraged by poor drainage, nodulation discouraged by wet conditions.	low	low—except in wet autumns.	medium

Crop					
Vining peas	medium	medium—disease encouraged by poor drainage, nodulation discouraged by wet conditions.	medium—for early drilled fields.	medium—although harvested at time when soil is normally dry, if there is a wet period damage to soil by heavy harvesters severe.	medium
Grain maize	medium	low—normally growing during drier season of year.	low—late seedbed preparation.	medium—harvested later than cereals.	medium
Brussels sprouts and winter greens	med/high	high—at risk during wettest part of year.	low—if transplanted moderate—if direct drilled	high—where machines and trailers are needed on land.	high
Summer cabbage and autumn cauliflower	medium	low—normally growing during drier period of year.	low—if transplanted moderate—if direct drilled.	low	medium
Bulbs	high	high—tulips medium—daffodils	high—deep tilth free from clods needed	medium—normally harvested in drier period of year.	very high
Dessert apples Sweet cherries Raspberries Gooseberries	high/very high	very high—at risk during wettest parts of year and roots easily killed.	low—frequent spraying needs reasonable land bearing capacity.	low	very high
Strawberries	high	medium	high—when runners are planted.	low	high
Blackcurrants	high	medium	high—when bushes are planted.	low	medium

NB—Crops with very high or high overall requirement for drainage normally sited on naturally well drained soils.

The single or double bladed subsoiler is another implement used to give permeability in tight soils. Compared with moling, the emphasis is more on lifting and shattering compact subsoil rather than on forming a stable drainage channel for conducting water. Consequently, the ground needs to be drier than for moling, and drains need to be closer if water cannot permeate the deeper subsoil. To some extent moling and subsoiling are inter-changeable, but subsoiling as a drainage aid should be reserved for clay soils with subsoils of variable texture, and for soils with a definite overcompact layer that needs shattering. More information is given in Chapter 13.

PLASTIC PIPE DRAINS

Long lengths of perforated plastic pipe can be used successfully in place of clay pipes, where local contractors have the necessary equipment. An advantage of plastic is that a narrower trench is excavated and so disturbance of the land is less; also the amount of permeable fill needed is reduced. Some contractors have equip-ment which lays plastic into a channel formed by a type of large mole blade, which leaves only a slit through the ground. Conse-quently, where drainage is put through growing crops, soil disturbance and crop reduction are minimised.

Although length for length plastics are cheaper than clay pipes, the equipment for laying it is more expensive. In addition, a 50 mm plastic drain takes less than half the water of a conventional 75 mm drain, and so designs based on plastics often need closer spacing if the laterals are long. For these two reasons plastics are often only slightly cheaper than clay pipe systems.

MAINTENANCE AND AFTERCARE OF DRAINAGE SYSTEMS

An intensive under-drainage system is an expensive investment that must be serviced regularly to get the best from it, and yet on many farms ditches are allowed to silt up and outfalls block, thereby endangering the whole system. Weed growth in ditches encourages silting up and should be cleared regularly; obstruction in the form of fertiliser sacks and other rubbish should also be regularly removed. Slipping banks need to be restored before under-cutting and erosion takes place, and in unstable sandy subsoils bank gradients may have to be reduced.

Blocked drain outlets are very common on farms and of course seriously impair the efficiency of schemes. The outlet should be marked with stakes at the bank top and regular inspections made to ensure they are clear. Frequently when ditches are cut, deepened, or opened up for the first time, outlets of old pipe drains are unearthed, which should be given new outlets if they still run. In some cases restoration of old systems with new ditches, and rodding of drains, can dry up a wet field without further expenditure. On land where pipes tend to silt up, regular cleaning of silt traps and rodding of drains to keep them open is well worth while.

It is important to realise that pipes are passive receptors of drainage water and are unable to suck water through any obstruction to water movement in the soil above. A drainage system is not therefore an insurance against working soil in all conditions, but an aid to maintaining better soil conditions by careful soil management. Compaction caused by machinery should be removed as soon as conditions are right, and the soil may need regular moling or deep subsoiling to maintain permeability and a good environment for deep rooting.

DRAINAGE 'WORTHWHILENESS'

There is no doubt at all that when symptoms of poor drainage are obvious and have been noticed over several years, money spent on a suitable drainage system gives good returns in better timeliness of cultivations, easier management and increased crops. However, in each case the question must be asked: how much money per hectare should be spent to give the best investment? With fruit, many vegetable crops, and root crops, there is little doubt that highly efficient systems are needed, but for lower value enterprises such as intensive cereals, beef and sheep, the situation is much less straight-forward. Unfortunately these enterprises are often found on difficult clay soils that need an expensive drainage system for efficient water control.

Farmers faced with large acreages of land requiring drainage, and having limited capital resources, need to decide whether it is better to drain a small area annually with an expensive but technically optimum system, or whether to tackle a large acreage annually with a much less intensive and technically less effective scheme with the opportunity (in the latter case) of returning in later years and filling in with more drains if needed.

On clay land, provided certain conditions are met, the right decision will normally be to drain more hectares less intensively, and to rely more on relatively cheap mole drainage, and less on expensive pipe drainage. The conditions which have to be satisfied are:

(a) Widely spaced tile drains must be covered with permeable backfill to remove water from moles.

(b) Pipe drains should be so positioned that they can, if necessary, form an integral part of a closer system at a later date, and will need to be closer on land with subsoils that are not ideal for moling.

(c) Moles should not be pulled through low areas untapped by pipe mains; otherwise wet areas will develop and increase in size.

(d) Moles should be pulled when soil conditions are as good as possible—through cereals in May-June is a very suitable period.

(e) Moles will usually have to be pulled regularly to offset their shorter expected life span, and remoling should be done before signs of deterioration appear. This can only be done by digging down to examine channels or by observing how soon field drains cease to discharge water after rain.

Provided farmers comply with these conditions, large improvements can be made with only small expenditure. For example, on a London Clay, where the optimum system is pipe drains 20–40 m apart with moles, a large measure of improvement can be achieved with a system 60–80 m apart with moles, although of course risk of breakdown in the mole channel is greater.

The ultimate in cheap drainage is to rely solely on mole channels. Provided fields slope uniformly without low pockets, and the subsoil is a good clay, effective drainage can be achieved by pulling mole drains over a moled main drain. The moled main should open into a ditch and be pulled parallel and close to the bottom boundary of the field. It should be re-drawn at least every third year.

CHAPTER 6

SOIL STRUCTURE AND CROP PERFORMANCE

The final yield of a crop is the result of all those factors which influence growth during the season. One of these factors, the physical condition of the soil, often has an important influence on crops because it controls the environment in which roots develop. Soil structure is the term normally used to describe this property of the soil. If soil is likened to a building, then bricks and mortar are equivalent to particles of sand, silt and clay, i.e. to soil texture, and rooms and corridors are equivalent to the units into which these particles are combined, i.e. the aggregates or structures.

Soil physical fertility, which we often attempt to define and measure as structure, involves five agriculturally important properties:

- Rainfall acceptance by the soil and the ease with which excess water is shed from the soil.
- Storage of 'available' water and the ease with which roots can retrieve this water.
- Mineralisation of organic matter in soil, which involves temperature, oxygen supply and moisture supply.
- Seed germination and early population establishment—influenced by packing of soil around the seed and interaction between soil temperature, moisture and pest and disease.
- Crop growth—dependent on root penetration and extraction of water and nutrients in phase with development needs of the crop.

STRUCTURE AND POROSITY

The fundamental importance of structure to farming is that it determines the level of compaction, both in the plough layer and below the subsoil. The solid structures of peds are less important than the spaces between and within them, i.e. the soil porosity. The systems of pores and fissures, which extend through the whole depth of the soil profile, control the ease of air and water movement, the quantity of available water that can be held, and the ease of root penetration and exploration.

The human eye can just detect objects about $\frac{1}{5}$ mm across. It is pores of this size and larger which allow drainage water to pass and roots to penetrate, and so, in general, any pores that can be seen by eye will permit drainage and root exploration, provided they are continuous. Soil also contains a network of smaller pores which, although they cannot be seen, are vital for the storage of water available to plant roots. Unless a soil profile has a good network of visible pores and fissures, drainage and aeration are likely to be inadequate, and the nutrients and water located in the finer pores are largely inaccessible to the plant roots.

STRUCTURE AND ROOTS

It is not always realised that if crops are to produce high yields, roots need to explore to a depth of 1 metre or more. For example, in good winter wheat crops, roots are usually 600 mm down by April, and in one case the authors found roots at 1.4 m in April in a brickearth soil. On similar soils sugar beet regularly extract water from below 1.2 m. The importance of an adequate root system for water uptake cannot be overstressed. Most of the water removed from the soil by roots is lost from the foliage by transpiration, and on a windy sunny day in summer about 3 mm of water (30 tonne/hectare) is transpired from a full crop cover. Unless the soil can supply this water, growth rate declines, and eventually if this continues yield is impaired. Irrigated crops are less dependent on deep rooting.

The natural water-holding capacity of soils obviously has a major bearing on yield potential in areas where water consumption by crops exceeds rainfall during the growing season. In addition, an artificial restriction to rooting caused by lack of consolidation, or overcompaction and anaerobic conditions, will accentuate any moisture deficiency. For example over eastern and southern England, the average potential soil moisture deficit (excess of evapo-transpiration over rainfall) is about 150 mm rising to 225 mm in a very dry summer. If sugar beet in a loam soil cannot root below 450 mm because of pans, then the maximum moisture that the crop can remove from the soil will be about 75 mm, or only enough to see the crop through 3 weeks of dry sunny weather before wilting occurs. If, however, the crop has rooted freely to 1.2 m, wilting will only occur after long periods of drought, and the risk of losing yield will be less.

TYPES OF STRUCTURE

In light sandy soils, structure is only weakly developed, and the porosity of the soil depends largely on the size of the sand particles and the closeness of packing. Clay soils, on the other hand, are completely lacking in porosity and permeability unless structure is present. Soils with intermediate textures such as sandy loams, loams, and silts have structures intermediate between the extremes of sands and clays. Soil organic matter plays a major role in promoting aggregation of light soils, and gives stability in medium textured soils and mellowness and friability to heavy soils.

Structures commonly found in soils are illustrated in fig. 6.1. In topsoils, where the agencies of change have their greatest influence, structures are usually a mixture of artificially produced clods and weathered granular and blocky structures. After a wet season, if the soils become compacted, the topsoil may consist mainly of very large compact or massive structures, which are only gradually broken down again. These large clods are sometimes so dense that roots can only squeeze between them, and any water or nutrients inside them are inaccessible to roots.

Structures in subsoils are more permanent and usually larger than in the plough layer. They vary from large prisms 150 mm across and 300 mm deep in some heavy clays, to small blocks 15 mm across in well structured clays and lighter soils. However, the important factor is not the size of a structure, but how porous it is and how closely it fits against its neighbours when a soil is moist and swollen.

FORMATION

The formation of soil structure is brought about by wetting and drying, by freezing and thawing, and by cultivations. In addition, the growth of roots, the activities of earthworms, bacteria and

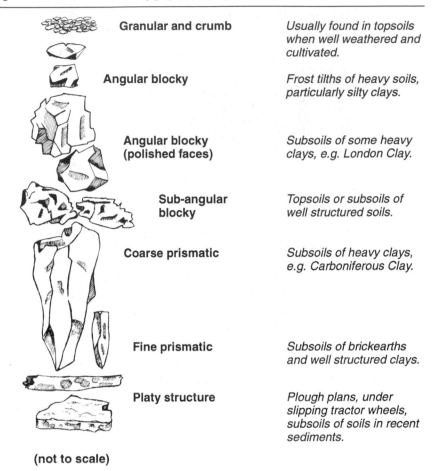

Granular and crumb	*Usually found in topsoils when well weathered and cultivated.*
Angular blocky	*Frost tilths of heavy soils, particularly silty clays.*
Angular blocky (polished faces)	*Subsoils of some heavy clays, e.g. London Clay.*
Sub-angular blocky	*Topsoils or subsoils of well structured soils.*
Coarse prismatic	*Subsoils of heavy clays, e.g. Carboniferous Clay.*
Fine prismatic	*Subsoils of brickearths and well structured clays.*
Platy structure	*Plough plans, under slipping tractor wheels, subsoils of soils in recent sediments.*

(not to scale)

Figure 6.1   Description of structures found in arable soils

other soil organisms also encourage structure. By these factors soil particles are joined together, or broken down into distinct units, their size and shape depending mostly on the type of soil, but also on the system of management. Wetting and drying cycles occur throughout the soil profile, and in soils of medium to high clay content the strains caused by swelling and shrinkage are a major factor in producing structural units (photo 6.1). In this country frost effects are largely confined to the topsoil but are vital for the management of heavy arable soils.

Roots grow through the soil, opening up fissures and exerting pressures both mechanically by their growth and through the

Plate 6.1   Deep cracking on clays in dry years helps to regenerate structure and improve drainage   (G.W. Cooke)

drying of soil around them. The fine granular and crumb structures, which are particularly well developed under old pasture, owe much of their fineness and porosity to the development of abundant fibrous roots. Cultivations play an important role in modifying topsoil structure, usually by loosening or consolidating, and by dividing larger units into smaller peds suitable for seedbeds. However, the formation of large units or clods by compaction and puddling under wheels, and compacting implements, is commonplace and a harmful influence on soil fertility.

STABILITY OF SOIL STRUCTURE

Unstable soils are ones which lose their structure easily when rain falls on the surface or water is held up in soil. In UK soils the major factors which impart stability are clay, soil organic matter and lime. Sometimes oxides of iron and aluminium are also important. In soils containing too little of these materials surface capping, slaking, and panning under wheels are more likely to occur. Instability is a feature of most soils containing appreciable amounts of sand and silt sized particles, particularly those with a low level of organic matter. The textures most likely to be affected by problems of instability are sands, fine sandy loams, silt loams, silty clay loams and sandy clay loams.

Wherever drainage is slow or there is a high water table, instability is much more evident, and the need for good drainage control is a high priority. Although stability increases as clay content increases, in practice the worst problems arise in soils with enough clay to give problems with clods and working properties, but not sufficient to give stability. This situation occurs in some silty clay loams and sandy clay loams of eastern England. Problems of managing unstable soils are considered in Chapter 15.

Clay-rich parent materials well endowed with finely divided calcium carbonate (lime), e.g. chalky boulder clay, in general give rise to soils which have better developed structure, and work more easily than non-calcareous clay soils, e.g. most Oxford and Carboniferous Clays.

TYPES OF POOR STRUCTURE AND THEIR OCCURRENCE

In Topsoils
Structural problems in topsoils arise in a wide range of soils, but are more common in those with slow water movement and unstable structure. In practice, therefore, more troubles are found on clays, silts and sands, and fewer problems on loams, sandy loams and chalk and limestone soils. The more cultivation land receives, the more common are structure problems.

Five types of problem arise which may affect plant growth:

1. *Surface capping.* Rain falling on to the fine surface of an unstable structured soil destroys the aggregates creating a close packed cap at the surface. This cap may vary between 1 mm and 40 mm in thickness, and can cause low plant populations particularly if it dries out before emergence (photo 6.2).

2. *Underconsolidation.* If tilth is too loose, drilling is likely to be too deep, and the seedbed dries out very readily. These conditions regularly occur in sands, light loams and peat soils. Where land is underconsolidated, better growth frequently shows in wheelings

Plate 6.2   Emergence of small seeded crops is jeopardised by capping on soils of unstable structure

(see page 212). However, with sands it is very easy to convert an underconsolidated tilth into an overcompacted one.

3. *Compaction and cloddiness* (colour photos 11, 12 & 13). Massive overcompact layers or zones may occur at any depth in the topsoil, and in the extreme all but the top 25 mm is involved. These structureless conditions are caused by compacting implements, and wheels of tractors, trailers, harvesters, etc. (photo 6.3). Sometimes compact layers have pronounced platy structure which is induced by slipping wheels. Very thin compact layers, caused, for example, by smearing as planter shares move through wet soil, can be sufficient to prevent root penetration. The influence of soil moisture content on soil compaction is covered in Chapter 8.

A soil with good structure may have 60 per cent of its volume as pore space, of which 20–30 per cent is occupied by air at field capacity—that is, when water has just ceased to drain. An over-compacted layer may have a total pore space of only 30–40 per

Plate 6.3   A shallow pan caused by smearing of wet soil during seedbed cultivation

cent, with as little as 5 per cent or less of the pores filled by air at field capacity. The loss of larger pores not only restricts movements of air, but also greatly reduces the rate at which soil drains. In addition, compact soil presents greater resistance to roots, and those that do penetrate are confined to fissures between structureless clods.

4. *Anaerobic layers* (colour photo 17). A particularly harmful condition is produced by inadequate aeration in soil. When fresh organic materials such as beet tops, straw or farmyard manure are ploughed under in wet compact soil, the need for oxygen outstrips supply and anaerobic conditions form. In severe cases the soil colour changes to grey and a foul smell of sulphides can be detected. Roots cannot live in these zones and crop loss may be severe if the anaerobic layer is continuous. Fortunately the condition is reversible and when air is introduced into the layer, by cultivations or drying, the grey colour disappears.

5. Slaked and puddled structure. In wet conditions the plough layers of arable soils sometimes lose their aggregated structure without becoming compact. Individual aggregates lose their identity as soil particles flow from them to fill voids and to meet similar material moving from neighbouring aggregates. In this slaked or puddled state (see page 117), water movement is very slow and the soil retains a large amount of water after the soil has ceased to drain any further water. Puddled soil will only bear light weights, and cultivations have to be delayed because they cause smearing and bring up large lumps of soil which dries into hard clods.

Unstable structured soils with moderate clay content are most vulnerable to slaking or puddling, and the condition is very much more common in overcultivated soils in wet years. Unstable soils are sometimes slaked by water alone, particularly if the water is held up over a compacted layer, but mechanical force from wheels and cultivation implements is the most frequent cause of puddling in soils (photo 6.4). Structure collapse is considered further in the section on deterioration of structure (see page 83).

Plate 6.4   Lifting root crops in wet autumns causes severe structural damage which may take several years to put right   (Crown copyright)

In Subsoils

For an account of adverse structures which occur naturally in subsoils in this country, readers should refer to page 186. Over-compaction below the plough layer is common in many soils, and there seems little doubt that ploughing with wheeled tractors in the furrow bottom is the main cause. Weak structured soils normally have narrower fissures between structures, and consequently plough pans form more easily in these soils. Excessive tractor wheel slip increases the risk of smearing and compaction (colour photo 16). Sands, gravels, silts, and clay soils are most susceptible to subsoil compaction, and although normally the compact layer is no more than 50–150 mm thick, in sands and gravels compaction may extend to 450 mm or more (see photos 6.5 and 6.6).

Plate 6.5   Oxford clay profile showing over-compact topsoil, plough pan and naturally compact blocky structured subsoil which drains very slowly   (Crown copyright)

Plate 6.6   Severe plough pan which caused premature drought in a Somerset orchard   (Crown copyright)

DETERIORATION OF STRUCTURE

It is important to realise that short-term visual changes in structure, for example compaction under wheels and shattering by frost, often mask very much slower changes in structural properties continuing over several years. These long-term changes may have important and far-reaching consequences for the farmer, but because they are less rapid they are less easily noticed.

The most widespread long-term change is the gradual deterioration of structural properties, with declining organic matter, in land ploughed from pasture as long as 30 years before. This process is particularly noticeable in clay and silt soils where it is accompanied by a general decline in the working properties of the plough layer (colour photo 11). The mellow friable consistency of the soil is gradually replaced by a harsher structure, with more angular frost tilth, and a narrow moisture range over which cultivations can be successfully achieved. Extra power is needed to plough the land, the influence of adverse weather is more harmful, and smearing and compaction more commonplace.

Another example of longer term change is the gradual decline in structure as the drainage system of a soil silts up or moles collapse. Risk of soil damage in the spring increases, and crops become poorer in patches which gradually extend in area. Eventually crop failures may occur, and the land becomes very difficult to work. This cycle of deterioration can be reversed by renewing the drainage system and opening up the soil with a subsoiler or mole plough.

In general soil compaction and puddling are reversible. Dry weather, frost and remedial cultivations will, in time, restore the structure of damaged soil, but, as many farmers know to their cost,

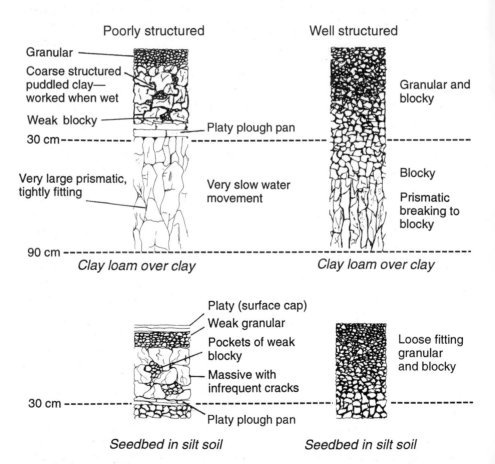

Figure 6.2   Poorly structured and well structured profiles

damage inflicted on difficult land by a sequence of wet autumns and springs can take several years to recover from fully. After a sequence of dry years, structure in any well farmed soil is usually in good condition, but it is the ease with which structure deteriorates in wet years that is important, and which distinguishes weak from stable structured soils.

IDENTIFICATION OF POOR SOIL STRUCTURE

With experience and careful observation of the soil profile it is possible to assess accurately the type and extent of any physical problems in the profile (fig. 6.2, colour photos 14 & 15). But first much useful information can often be gained from the surface features of the field, for example weedy areas and variation in crop vigour. Having selected typical sites, preferably in areas of contrasting crop growth, the first step is to examine an exposed vertical face of the soil in the side of a pit. The pit need be no deeper than 450 mm if the aim is to locate damage caused by machinery, but in cases where the natural properties and potential of the soil are of interest, deeper pits of 1–1.5 m are needed. The ease of digging can be a useful indication of compaction, but frequently healthier crops have dried out the soil more and make digging more difficult. For deeper pits a mechanical excavator is much more convenient.

The key features to look for are: changes in structure, colour, moisture, and rooting down the soil profile. The type of structure is best examined by removing spadefuls of earth from each depth at the side of the pit, and assessing the type of structure, its porosity, and how tightly the peds are packed. Overcompaction is identified by a relative absence of porosity and of large massive structures, or absence of structures altogether in extreme cases. Frequently colours are duller in compacted topsoil. The ease with which a walking stick can be pushed into the soil is a useful guide to the presence of pans. Dull grey colours and grey patches or mottles, rather than bright brown homogeneous colours, indicate lack of aeration and periodic waterlogging, which in wet times can be confirmed by the presence of water in the soil. Where there is a growing crop the root distribution is a useful key to the interpretation of structural features.

Without experience in examination of soil and interpretation of the findings, the significance of what is seen will not be understood by most farmers. In the first instance it will be useful to seek help from an experienced specialist, but the skills of soil examination can soon be acquired and eventually regular examination of the soil to assess the needs of, and success of, cultivations should be the aim of arable farmers.

STRUCTURE AND BIOLOGICAL PROCESSES IN SOIL

The importance of biological processes, and of soil organic matter, in the formation and stabilisation of structure, has already been mentioned in Chapter 1 and earlier in this chapter. Residual organic matter, or humus, plays a major role in binding together mineral particles of soil, and bacterial gums and fungal hyphae exhibit a more ephemeral role in stabilising aggregates. Although humus and other organic materials play a particularly useful role in soils with natural instability, the working properties of clay soils are also much improved by 1 per cent or more extra organic matter (see Chapter 14).

The overall biological activity of a soil often appears to be related to its capacity to grow good crops, and the evidence of partially decomposed plant remains, as long as two or three years after they were turned in, is a useful indication that the soil is not in 'good heart'. In these cases the soil structure is often poor, but whether deterioration in structure causes lower biological activity, or is the result of it, is not clear.

Darwin was particularly impressed with activities of earthworms in soils, and wide experience of agricultural soils today confirms the continued importance of these creatures in improving soil fertility. Even in intensively cultivated land, significant amounts of topsoil pass through the guts of worms annually and emerge with improved aggregation. In addition, earthworm channels penetrate deep into the subsoil and are often a major pathway for roots and water movement. Sometimes these channels are the only connection between topsoil and subsoil through solid plough pans.

SOIL STRUCTURE AND FERTILISER USE

Whenever a soil condition limits root exploration, nutrient uptake may be so much reduced that growth rate suffers. In practice only two nutrients—phosphate and nitrogen—are involved. When crops are grown in poorly drained or wet compacted soil, water soluble fertiliser phosphate should be placed with the seed, to ensure sufficient of the nutrient is taken up in the early stages. This is of course particularly important in soils low in phosphate.

Crops growing in poor soil conditions very commonly show signs of nitrogen deficiency, and extra fertiliser nitrogen often has to be applied. The deficiency is caused partly by reduced uptake from a poorer root system, but also because nitrate is lost by bacterial denitrification in anaerobic soil. Cereal crops showing deficiency from this cause usually need an extra 25–60 kg/hectare of nitrogen as a top dressing.

This extra requirement of crops for nitrogen is particularly noticeable in peas, which in normal conditions give best results without nitrogen, but in poor wet seedbeds can give a worthwhile response to applied nitrogen. Several examples of this effect were seen in Lincolnshire in the wet spring of 1969; in one case peas growing on a silt loam, badly compacted at 100 mm, were very much poorer one side of a line through the field. The only difference in treatment either side of the line was 40 kg/hectare nitrogen applied in the seedbed of the better peas as against only 14 kg/hectare in the poorer peas. Plants in the poorer areas had roots without nodules, showed symptoms of acute nitrogen deficiency, and were attacked by foot rot. The yield of green peas was only 450 kg/hectare in this area. The roots of plants in the better section had healthy nodules, no foot rot, and plants showed only slight symptoms of nitrogen deficiency. The yield of peas in this part was 2.5 tonne/hectare.

Frequently symptoms of magnesium deficiency occur in crops growing in adverse soil conditions. Magnesium fertilisers have little or no effect on the yield of these crops, though the colour of the crop may be improved.

IMPORTANCE OF OTHER FACTORS

It is impossible to predict accurately the effect of a given level of soil compaction on the yield of a crop. This unsatisfactory situation exists because of the variety of factors which modify the plants' response to a given soil condition. The most important of these is weather. In general a structural defect tends to have more effect on crops in unusually wet or dry conditions. For example, in a strongly panned topsoil, very wet weather may result in water-logging, and in a drought period restricted root penetration will cause prolonged wilting. The ideal weather for this condition would be a changeable period with periods of rain and sun. Again, in the case of a severe surface cap, a dry period after the cap has formed may completely prevent emergence, but if the cap remains moist emergence may be unhindered.

Frequently cases are noted in the field where disease and pests modify response to a soil condition; foot rot in peas, eelworms in potatoes, and take-all in wheat are the most common. In general, a soil structure problem aggravates the effect of a given level of disease or pest in the soil. For example, in the case of take-all in heavy soil, if the soil is in good condition, a fair level of disease can be tolerated without much effect on yield, but if compaction causes slower drainage the same level of disease will be much more harmful.

SOIL STRUCTURE, FARMING SYSTEM AND CROP SUSCEPTIBILITY

The optimum system for a farm depends on climate, soil, financial considerations and a farmer's skill with, and preference for, particular crops. It is important to choose crops which are compatible with the climate and soil, and ones which do not involve an unacceptably high risk of soil damage and crop loss. Often it is not simple to determine the risks involved, partly because of the variation in weather from year to year, but also because if there is a change from a grassland system to one dominated by arable cropping, a gradual decline in organic matter and soil working properties will take place over the next 20–30 years. However, despite the difficulties it is vital to make as accurate an analysis of the risks as possible, and experience of advisers, and neighbouring farmers on similar land, should be taken into account. The actual financial budgeting for a proposed change of system is easily accomplished, but the real crux of the problem is the accuracy with which budgeted costs and yields in the new system can be realised. The possible need for extra capital expenditure on drainage, in cases where a change is proposed to more exacting arable crops, should not be overlooked.

The sensitivity of different crops to soil structural conditions varies considerably. In addition, some crops are sensitive to one type of problem but not to others. Winter cereals are less sensitive to all problems than spring cereals, which have a shorter growing season and need better conditions for adequate tillering. Sugar beet and other small-seeded crops are particularly sensitive to seedbed conditions and to problems of early seedling develop- ment (photo 6.7), whereas potatoes are more affected by soil conditions restricting root exploration and water uptake in the mature plant. Peas and beans are sensitive to wet springs, when compaction in the topsoil reduces nodulation and encourages damage from soil-borne foot rots. Crops such as carrots, parsnips, sugar beet and brassicae are particularly sensitive to loss of the tap root (photo 6.8). If this root is lost because of compaction or anaerobic conditions, the depth of rooting is usually less, and in root crops the storage organs become fanged and yield less. With parsnips and carrots for prepacking, this damage results in lower graded produce and reduced profit.

Table 6.1 summarises the sensitivity of a number of common agricultural crops to several soil problems, and the risk of damage to soil from growing these crops.

Plate 6.7   Effect of poor soil structure in restricting growth of sugar beet (Crown copyright)

Table 6.1   Sensitivity of crops to specified soil conditions in the top 450 millimetres

Crop	Tilth requirement	Susceptibility to waterlogging and anaerobic conditions
Winter cereals	Requires moderately fine tilth for uniform establishment in dry autumns and for efficient performance of soil-applied herbicides.	Tolerant.
Spring cereals	Requires moderately fine tilth for uniform establishment.	Delays date of drilling, restricts tillering, and accentuates need for nitrogen.
Potatoes (maincrop)	A deep tilth free from clods of greater than 40 mm is required for machine lifting.	Sensitive to rot if ridges waterlogged for 24 hours.
Sugar beet	Requires fairly fine even seedbed but not too fine. Should not be panned below seedbed.	Very sensitive during germination and early establishment phase.
Peas	Needs moderately fine tilth.	Very susceptible in wet springs to attack from foot rots and nitrogen deficiency because of poor nodulation.
Field beans	Will establish well in coarse tilth. Winter beans can be satisfactorily sown by ploughing the seed under.	Sensitive to foot rots in wet springs.
Carrots and parsnips for prepacking	Fine tilth needed.	Very sensitive—reduced emergence and harmed tap root.
Brussels sprouts	Requires fairly fine even seedbed if direct drilled. If transplanted, fine tilth unnecessary, but loose conditions are not suitable.	Unlikely to occur during establishment.
Short-term leys	Fine even seedbed required.	Sown species readily lost, and weed grasses take over.

Requirements for other crops can be obtained by referring to a similar crop in this table.

Table 6.1 (continued)

Susceptibility to capping and compaction	Susceptibility to wind erosion	Risk of damage to soil
Fairly tolerant both to capping and compaction.	Very tolerant because of complete ground cover in spring.	Slight except in areas of high rainfall.
Tolerant to all but severe caps. Fairly susceptible to compaction which aggravates drought problems.	Tolerant because of early ground cover.	Heavy soils often compacted at seedbed preparation. Slight risk of damage at harvest except in areas of high rainfall.
Not sensitive to caps, severe compaction under ridge causes premature senescence.	Tolerant because ridges reduce wind velocity at surface and because of large seed.	High risk with late lifted crop because of heavy machinery.
Very sensitive to caps. Tap root sensitive to restriction in tight layers causing restricted water uptake and fanged beet.	Very sensitive because of fine seedbed and late ground cover.	Very high risk because of late lifting and heavy machinery.
Not sensitive to caps. Tap root sensitive to compaction.	Tolerant except for late drillings.	Dried peas only slight risk. Moderate risk for vining peas because of fixed harvest date.
Not sensitive to caps. Tap root sensitive to compaction.	Tolerant because of early ground cover.	Moderate risk because of late harvesting.
Very sensitive to even slight caps. Tap root very sensitive to compaction; size, shape and uniformity of produce badly affected.	Very sensitive because of fine seedbed and late ground cover.	High risk if lifted late.
Direct drilled crops very sensitive to cap. Roots badly restricted by compacted layers and plant fails to establish a large frame.	Not grown on soils where erosion is a risk.	High risk because of late harvesting.
Sensitive to capping. Depth or rooting easily restricted by compaction.	Not susceptible unless sown in late spring.	Risk of poaching damage.

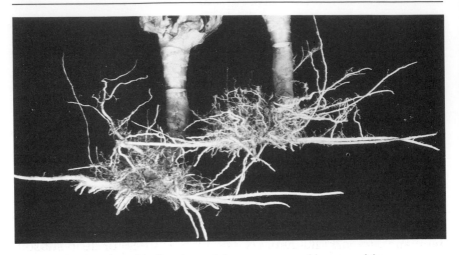

Plate 6.8   Root restriction in cabbage caused by machine planting on wet soil   (Crown copyright)

MAINTENANCE AND IMPROVEMENT OF STRUCTURE

Some soils and some cropping systems are more prone to soil structure problems than others. The worst problems occur on difficult land growing structure-sensitive crops, which also involve a high risk of soil damage, for example potatoes and sugar beet grown on unstable silty clay loams in the Fens. On the other hand, cereal growing on the chalk areas of southern England presents negligible soil problems. On difficult land, particularly if it is both naturally slow draining and weakly structured, the aim at all stages should be to keep the land in as satisfactory a condition as possible. In this way the extent of soil damage and crop loss is minimised in the event of long periods of wet weather.

Correct Choice of Crops

Many factors make up good soil management, but the first essential is to grow the right crops for the soil and climate. Of course the choice of crops must take account of financial factors as well but it is important to ensure, for example, that high gross returns from a crop like sugar beet or potatoes are evaluated in the context of higher labour and machinery costs, possibly additional expenditure on drainage, and risk of lower cereal yields after the beet. Again, it is important to take full account of the improved fertility, workability, and possibly higher yield of arable crops, that are associated with a grassland enterprise before eliminating it from the farm. An additional problem of changing from grass to arable is the financial difficulty of reversing the change at a later date. Some advantages and disadvantages of changes in farming system are obvious and easily measured in money terms; others may be equally important in the long term, but much more difficult to assess financially.

Good Drainage

The second requirement is adequate drainage, either natural or artificial. It is often not necessary to have a chain-apart system, but it is important to make sure that the existing drains are functioning well and regularly maintained, and where moling or deep subsoiling is part of the system, to make sure that these are repeated before the soil becomes more difficult to cultivate and yields decline.

Reduced and Timely Cultivations

Cultivations are the day-to-day means of providing optimum conditions for seed emergence and root development. However, besides their beneficial role, cultivations of the wrong type at the wrong time can do harm. Therefore good management must include awareness of the double-edged nature of this tool, and a real appreciation of what each crop needs and how this is best achieved. The first principle of cultivations is to do no more than is needed. Every extra cultivation means another set of wheelings and more expense. The greatest scope for saving is in cultivations for cereals, and there is now sufficient experimental evidence and proven practice to encourage all farmers to develop non-ploughing systems to suit their conditions at least for winter cereals. There is also great scope for reduced seedbed work with sugar beet and peas.

Many cereal farms now have the absolute minimum of labour, and inevitably this result in some cultivations not being done at the right time. Reduced cultivations and faster work are the obvious ways of achieving timeliness.

Timeliness of cultivations is an essential part both of drilling crops at the right time and of achieving good seedbed conditions. For example, for every week after mid-March that spring barley is drilled, yield on average falls 200 kg/hectare in eastern England; on the other hand, if the soil is too wet, and the seed is drilled

94 SOIL MANAGEMENT

into a poor cloddy seedbed, then yield can suffer more than if cultivations are delayed. The correct approach is to develop a system of preparing and drilling land in as short a time as possible once conditions are right; otherwise in a catchy spring some barley will be drilled far too late. In this context there is great scope for drilling direct into autumn-ploughed land using a tine drill. Cultivations for structure-sensitive crops such as sugar beet and potatoes should not be attempted if the soil is too wet, but once it is ready the output per day should be the maximum possible.

Wheel Slip and Crawler Tractors
The importance of tractor wheel slip as a source of compaction and damage to soil structure is discussed in Chapter 8. When wheel tractors are used on moist land they should carry enough ballast to give good traction, and the draught should not be so high that excessive wheel slip is inevitable. The advantages of crawler tractors for the maintenance of structure on clayland are well recognised, and provided the cost comparison is reasonable, and lack of flexibility not too important, structure can be more easily maintained with tracked vehicles.

Root Crops and Damage to Land
Whenever a field has been damaged by harvesting late crops in wet years, the aim should be to put things right as soon as conditions allow. This may not be possible before the next crop is harvested, but as soon as this is off, the soil should be examined, and the appropriate loosening cultivation carried out. Some fields on the farm are usually more readily damaged than others because of their soil type. It is worth while keeping root crops off these altogether, or if this is not possible, then lifting the crops as early as possible. These fields should be ploughed or cultivated early in the autumn before they become too wet.

Pests and Disease Problems
Some farming systems are much more prone to soil pest and disease damage than others. For example, when several winter wheat crops are grown in sequence, take-all and eyespot are more likely, or if peas and potatoes are grown in close sequence the risk of foot rot and nematode damage, respectively, is greater. Crops growing in land known to harbour pests or disease are more susceptible to attack if root systems are already poor. In these situations farmers should take particular care to avoid pans.

Grass and Farmyard Manure

Yields from intensive arable fields sometimes decline after punishment in a sequence of wet seasons, or because of a combination of soil borne disease and deteriorating structure. Repeated heavy dressings of farmyard manure, or well fertilised 2–3 year grass or lucerne leys, are often effective in improving this condition, provided the manure is readily available and the leys can be utilised. These remedies are particularly appropriate for weakly structured soils which are more prone to deterioration. However, grass roots are easily stopped by pans and compacted layers must be broken up before grass will grow satisfactorily.

Lime and Fertilisers

In acid soils biological activity declines and aggregation becomes poorer; consequently, good structure is encouraged by regular liming to maintain a neutral or alkaline reaction in the soil. Finely divided limes such as sugar beet factory lime or water works lime are particularly beneficial to structure and can be effective even where a soil is not acid. Dressings of 50–150 t/ha mixed thoroughly into the soil have given much improved structure on a range of soil textures including sandy clay loams, silty loams (brickearth), silty clays and clays. Because of the risk of contributing to trace element deficiencies, this practice is not advised on land already suffering from trace element deficiencies or on land where root crops are regularly grown.

In fields where crops are already suffering from bad structure, extra nitrogen will often partially compensate for the condition provided there is enough moisture in the soil for the roots to utilise the fertiliser.

In Chapter 15 there is further discussion of appropriate cultivations for maintaining structure and achieving timeliness for four soil types—sands, clays, silts and peats.

CHAPTER 7

SOILS AND TRACTION

The draught developed by a tractor depends largely on the shear strength* of the soil in contact with the tyre or track. Wet soil does not have much strength and so traction on wet soils is poor. Settled stubble land in the autumn is stronger than soil which has been cultivated and therefore gives better traction. When pressure is applied to soil, its strength is greater and so extra weight in a tractor gives extra draught.

COHESIVE AND FRICTIONAL SOILS

The shear strength of agricultural soil is derived from a combination of its cohesive strength, which can be compared with the resistance of well-worked putty to tearing, and its internal friction, which can be compared with the friction between two solid bodies such as a brick being drawn over a table.

Wet clays have no internal friction but have cohesion which is not dependent on the load applied. Pure sands derive their strength only from the internal friction which increases, as with the brick on the table, with the load applied to them. Agricultural soils do not fall simply into the either classification but are a mixture of each with sandy soils tending to respond to loading to increase their strength, and the strength of clay soils depending more on the area that can be sheared. In practical terms traction on sandy soils is most likely to be obtained by ballasting; while on clay soils traction is obtained by means of large contact areas, such as with a crawler.

---

* Shear strength is a measure of the force needed to shear soil and is expressed in SI units as kN/m^2. Some methods of measurement are mentioned in Chapter 10.

SLIP

The strength of the soil cannot be converted into draught without some displacement of soil by the tyre or track. The amount of displacement or slip for any one tractor and soil situation depends on the draught being developed (fig. 7.1). Full advantage of the strength of the soil can only be taken at a certain amount of slip, and an increase in slip beyond this point reduces the draught that can be developed. Tractors in the field should certainly never be operated anywhere near the level of slip at which draught declines, and compaction and smearing damage of the soil occurs well below this level.

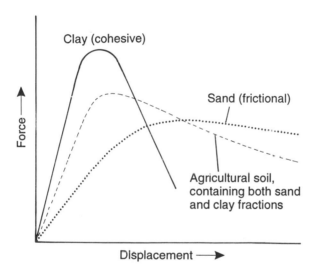

Figure 7.1   The relationship between force (or draught) and displacement (slip)

WHEEL LOADING

Increasing the load on a tractor tyre on practically all soils increases shear strength and therefore maximum gross draught (fig. 7.2). Gross draught is quoted because extra loading could result in increased rolling resistance through sinkage of the tyre and a reduced net draught available for useful work. In general it can be taken that the draught available is roughly proportional to

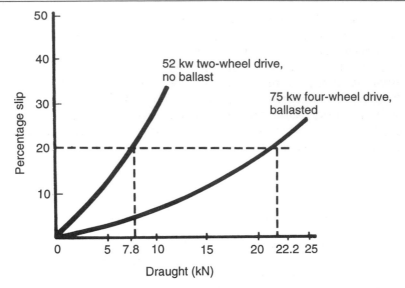

Figure 7.2   Performance of two tractors in the same clay stubble field on the same day. Note the differences in draught available at 20% slip level

the weight on the driven wheels, and that higher powered tractors must therefore be heavier in order to transmit extra power.

The proportion of the weight of a tractor which is applied to the driving wheels varies considerably. A two-wheel drive tractor without ballast and with no weight transfer may have 60 per cent, while the same tractor with a lot of ballast or weight transfer may have 90 per cent on the driven axle—in effect the minimum weight on the front end necessary to retain steering. A four-wheel drive tractor without weight transfer has 100 per cent of its weight on driven wheels, and with transfer from a heavy implement may have the equivalent of 120 per cent on driven wheels.

FORWARD SPEED AND WHEEL LOADING

Since draught developed depends on weight on the driven wheels, and power can be calculated from draught × speed, the weight required per kilowatt can be set out for a range of speeds for both types of tractor. Table 7.1 sets this out for good tractive conditions.

Table 7.1 makes the point that very heavy tractors would be required to make use of the power available if operating speeds were to be as low as 1.6 km/h, and that the weight required drops off quickly with increased speed. By the same reasoning, for a fixed weight of tractor less slip and probably less soil damage and mechanical inefficiency will be involved if higher speed (and lower draught) is employed to make use of the available power. These remarks do not of course apply to crawlers which are heavy for their horsepower but exert low ground pressures and, unless fitted with rubber tracks, are comparatively limited in their range of available speeds.

Table 7.1   Tractor weight and speed of operation

Field speed km/hr	Required weight kg/kW	
	2-wheel drive	4-wheel drive
1.6	268	189
6.4	67	47
9.7	44	31
13.0	33	24

Source: Institution of Mechanical Engineers.

The following are some of the terms commonly used in discussing traction:

Drawbar pull: Defined as the horizontal force developed by the wheel or track. It depends on the gross effort developed from the power applied to the wheel or track and the losses involved in moving it through the soil. The difference between the two is the drawbar pull available for useful work. Drawbar pull, for the reason already stated, is usually quoted with a corresponding slip value.

Coefficient of traction: Defined as drawbar pull divided by wheel load. Thus for a given coefficient of traction pull will increase in proportion to the load on the driving wheels. It is a measure of a certain soil condition or situation in relation to the tractive performance of a given tractor.

Tractive efficiency: Defined as the ratio of drawbar power to the power input to the wheel. The difference between the two is made up of the rolling resistance of the tyre and slip.

Force efficiency: Similar to tractive efficiency, but defined in torque input and output terms.

Slip: Stated as a percentage reduction in travel distance at the given loading compared with the situation where there is no slip. For practical purposes the no-slip distance can be taken to be the distance travelled by the tractor with no load in a given number (7 is usually convenient at about 30 metre run) of revolutions. The same number of revolutions under load will give the travel distance lost by subtraction and the percentage loss can be calculated as follows:

$$\text{Percentage Slip} = \frac{\text{Distance for 7 revs no load} - \text{distance 7 revs loaded} \times 100}{\text{Distance for 7 revs no load}}$$

It might appear that in a sandy soil, where traction depends largely on friction, a smooth tyre properly weighted would develop the full thrust required. In practice this is not so because the loose surface layer has to be cleared to allow contact with firm soil beneath.

TYRE LUGS

The functions of tyre lugs are:

- to sweep away loose soil and trash so as to expose firm soil beneath—some wheel slip is required to give this action and therefore some loss of efficiency is involved;
- to penetrate a water film on concrete or a slippery layer on hard cohesive soils; and
- to provide a rack and pinion effect on cohesive soils (clays) which are soft enough for the bars to penetrate.

The height of the lug has little effect on traction provided there is enough depth of tread to get through to firm soil. Very high lugs give reduced efficiency and greatly increased wear on hard surfaces, and the chief advantage of high lugs for farm work is that they have more to wear, and therefore promise longer life, than a lower lug. Worn tyres may be an advantage when tractors have to be used in growing crops because the sweeping and scuffing action of the lugs is less severe. Lugged traction tyres should not be used on tractors intended mainly for road haulage work. Tyres designed for this purpose will give greater efficiency and economy. Fig. 7.3 shows how tyre lugs wear on a hard surface.

Lug pitch (the centre to centre distance between lugs) and lug angle are chiefly fixed by the need to get good self-cleaning as well as good traction. The most common angle is 45°, that being reduced to $22\frac{1}{2}°$ in some cases. To maintain flexibility of the tyre the lugs do not cross the entire width; this is the 'open centre' design universally used for land work.

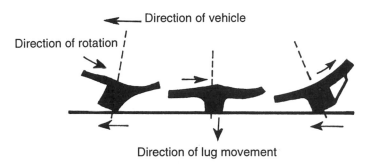

Direction of vehicle

Direction of rotation

Direction of lug movement

Figure 7.3   Illustrating the reason for severe lug wear on hard surfaces: the 'heel and toe' wear pattern (IMechE)

Tyre Profile

Tyre size is generally quoted as section width and rim diameter in that order. The designation 12.4-36 therefore refers to (1) and (3) in fig. 7.4, and replaces 12.4/11-36, where '11' referred to an out-dated nominal section width. Imperial units have generally been retained for traction tyres, but there are exceptions with mixed metric and imperial numbering. For example, 600-26.5 is a commonly used traction tyre quoted in millimetres and inches respectively. Large flotation tyres may give overall diameter (4) before section width and rim diameter as in 66 × 44.00-25.

In the UK tyre diameter increases width section for any one rim diameter. In Table 7.2, for example, tyre diameter increases from 58.2 to 60 in on the 36 in diameter rim as section width increases from 12.4 to 13.6 in. Most tyre sizes have a recommended rim width for optimum performance, and a slightly narrower permitted rim width. One tyre size wider can therefore usually be fitted if necessary.

The aspect ratio of the tyre is the ratio of section height (5) to section width (1) and for traction tyres is usually 0.85 or 85 per cent. In some cases it may be much lower and may be indicated on the tyre in the form 600/55-26.5. This indicates 600 mm section width, 55 per cent aspect ratio and 26.5 in rim diameter. A given tyre can usually be fitted on to a wider rim than that originally

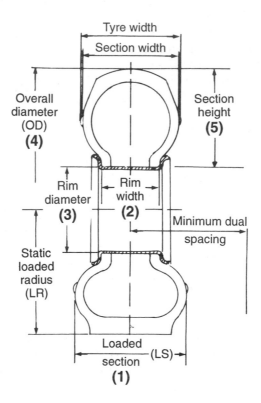

Figure 7.4   Tyre and rim description. A loaded tyre is depicted so that overall diameter (4) will be twice the distance from the centre line to the highest point. The catalogue 'OD' figure always refers to the 'free' or unloaded tyre   (Goodyear Tyre Co.)

Table 7.2   Tyre size and inflation pressure for a given load

	Width (in)	Diameter (in)	Load (kg)	Inflation (kPa)
12.4-36	12.4	58.2	1,340	152
13.6-28	13.6	51.8	1,340	138
13.6-36	13.6	60.0	1,340	110
13.6-38	13.6	62.0	1,340	103
14.9-30	14.9	56.0	1,340	97

Source: Institution of Mechanical Engineers

specified to give it a wider tyre section. The advantages of wide sections are slightly better stability, the ability to carry a greater load at the same air pressure or the same load at lower pressure, and greater volume for liquid ballast.

For the same tyre diameter neither section width nor height appears to have much effect on coefficient of traction or tyre efficiency within the 11–18 in rim range.

Inflation and Tyre Deflection

The thrust developed by the reaction of the tyre lugs with the soil results in some distortion or deflection of the tyre, and this deflection increases with increased load and reduced tyre pressures. Since tractor-operating speeds are low compared with road vehicles, the amount of deflection allowed is high at 18–20 per cent. Table 7.2 illustrates the point that if a given load has to be carried at reduced air pressure, tyre size has to be increased.

Manufacturers' catalogues set out acceptable loads and tyre pressures for each tyre size generally at 30 km/hr, with certain overload exceptions for low-speed use. There are radial tyres available for speeds of 50 km/hr, and for specialised tractors up to 90 km/hr. Since deflection represents strain on the tyre wall and only 9–10 per cent of the recommended load is carried by the actual carcass stiffness of a crossply tyre, it is desirable that correct tyre pressures are used. It is known from service records that the largest single cause of tyre failure is over-deflection caused by under-inflation.

Reduced tyre pressure also gives a higher rolling resistance and therefore reduced efficiency for road work. On soft soils and where sinkage is considerable, however, increases in drawbar pull of 5–25 per cent have been recorded for a given percentage slip. Again, it must be stressed that reduced tyre pressures can be expensive in terms of reduced tyre life.

Water-ballasted tyres can be run at higher deflections than air-filled tyres without problems of wrinkling, and wider rims for a given tyre again tend to reduce wrinkling.

Over-inflation prevents full contact of the tyre with the ground and excessive wear occurs at the centre of the tread. The tyre is also less able to withstand shocks, and localised damage to the casing may result.

Ply Rating

Ply rating is an index of tyre strength and does not necessarily refer to the actual number of plies built into the tyre. Using extra-ply tyres does not give longer life at a given pressure (and in fact may give a shorter life at low pressures), but allows higher inflation pressure and therefore heavier loading for higher power transmission and less wall damage under these high loads. 10-ply rating tyres are only likely to be advantageous on very heavy tractors. Radial tyre sidewall markings generally carry load indices to replace ply-ratings, and speed symbol letters at which the load index applies.

TYRE CONSTRUCTION

The ideal tyre carcass would be flexible and yet would not wrinkle under high driving torques, would carry heavy loads and would be resistant to external damage and immune to fatigue. The compromise answer to these conflicting aims is a tyre built up of layers of cords, or plies, set in rubber. Figs. 7.5 and 7.6 show types of tyre construction, the radial-ply being widely favoured for tractor driving tyres because of its excellent wear characteristics.

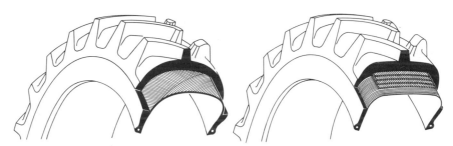

Figure 7.5   Crossply tyre construction. The casing is made up of several crossed plies. The crown is not stabilised (Michelin Tyre Co. Ltd.)

Figure 7.6   Radial-ply tyre construction. The belt is made up of several plies to stabilise the crown. There are one or two radial casing plies   (Michelin Tyre Co. Ltd.)

Crossply
In the crossply the individual cords lie at an angle to the centre line of the tyre and each layer of cords is set at the same angle but in the opposite direction to its immediate neighbour. This design of tyre is considered especially suitable for agricultural work because of its good resistance to wall damage, resistance to high impact loading and good crown flexibility which is an asset in wet clay soils.

Radial-ply

Radial-ply tyres have the cords laid directly across the tyres from side to side. There is in addition a belt of plies, laid at cord angles of about 20° to the central line, around the crown of the main casing. This serves to brace the tread and lengthen the life of the tyre. The performance of radial-ply tyres is better than crossplies under certain conditions, notably on light dry soils, and they can generally be expected to give 5–8 per cent higher pull in the normal operating range of slip (up to 20 per cent) than crossplies. The advantage, however, depends on using the lowest permissible inflation pressure since performance of the two types of tyre is approximately equal at higher inflation pressures. Flotation (resistance to sinkage) characteristics are better with the radial tyre. The length of life of radials is, accidental damage excepted, much longer than crossplies.

TRACTION AIDS

Ballast

Any increase in load on the driven wheels (photos 7.1 and 7.2), provided it does not cause excessive sinkage, leads to increased drawbar pull and decreased slip. The effect of liquid and cast-iron ballast is similar provided that the wheel load and inflation pressures are the same. While cast-iron weighting is most

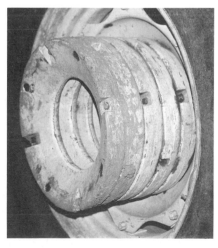

Plate 7.1   Rear wheel weights   Plate 7.2   Front-end weights

convenient in being easy to fit and remove in a series of stages, it is expensive and weights are generally not interchangeable between makes and models of tractors. Water ballast is generally used either as 80 per cent or 100 per cent tyre filling. The latter requires special equipment, and any water ballast, once installed, is unlikely to be removed to match the type of work and traction conditions in hand. Water ballast usually has calcium chloride added as an antifreeze.

Correct ballasting is important if traction efficiency is to be maximised and soil damage caused by wheel slip minimised. Research at Silsoe Research Institute indicates that maximum efficiency occurs at about 12 per cent wheel slip (less than the level of slip normally detectable by eye) and that loading equivalent to 0.9 kN for each kW of engine power is required on the driving tyres when the forward speed is 2 m/s (this is equivalent to 150 lb per hp at $4\frac{1}{2}$ mph). The loading required per kW declines rapidly as speed is increased, indicating that where possible higher forward speeds and reduced implement sizes should be selected for efficient tractor operation.

Most two-wheel drive tractors need considerable extra ballasting to give the loading indicated above, and larger size tyres or duals are often required. Four-wheel drive tractors generally have adequate tyre sizes fitted as standard.

Cage Wheels

Cage wheels effectively reduce ground pressure and are therefore an advantage where sinkage and soil damage is a problem (fig. 7.7). Under spring seedbed conditions, where cages are likely to be used to reduce ground pressures, they do not generally give increased drawbar pull and can be expected to give reduced efficiency since rolling resistance is high. There are some conditions where cage wheels can key into moist cohesive soil and give considerable increase in drawbar pull. An example of this would be for the first cultivation over ploughed heavy land.

Cage wheels are usually fitted smaller than tyre diameter so that they can be used on hard surfaces without damaging them. They are more likely to be effective in reducing ground pressures, however, if they are fitted at the same diameter as the tyre. Quick-release devices are valuable.

Strakes

Strakes dig into the soil beside the tyre and thereby make use of its cohesive strength. In clay they can give draught improvements of

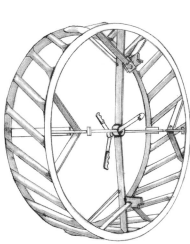

Figure 7.7 Cage wheel
(F.A. Standen & Sons)

Figure 7.8 Steel skeleton wheel
(F.A. Standen & Sons)

up to 200 per cent but have little or no effect in frictional (sand) soils. Strakes increase the rolling resistance of the tractor considerably, and therefore lower its efficiency. They should always be retractable and only extended in work to the minimum necessary to get the required traction or additional stability on slopes.

Steel Wheels

Steel wheels with spade lugs again make use of cohesive soil strength. They are less efficient than rubber tyres but can provide higher draughts under difficult conditions in clay soils. They can also be useful where flints cause such damage to rubber tyres as to make them very expensive for heavy draught work.

Because steel wheels give a positive key into heavy soils they may set up excessive strain in the tractor transmission and cause damage. Some tractor warranties are not valid if steel wheels are used.

Steel skeleton wheels (fig. 7.8) provide higher draught in heavy soils than either spade lugs or rubber tyres, but have a higher rolling resistance and, because of their lack of contact area to give flotation, tend to sink when implements are raised on the hydraulic linkage.

Tyre Tracks

Tyre tracks are available for fitting over standard rear tyres and additional tension tyres (photo 7.3).

They give an average ground pressure lower than that of conventional agricultural crawlers, give exceptional traction under difficult conditions and can be quickly taken off. Problems are the damage to track and tyre caused by sharp stones, wear on the tractor tyre inside the track if the tyre is nearly new, and a tendency for the tyre to run out of the track if large sideways forces are involved. There may also be problems in getting the tracks into plough furrows with the larger sizes since the track is necessarily much wider than the tyre it covers. This device is most successful on stone-free soils, and is best used over tyres that are already well worn.

Plate 7.3   Rubber tracks fitted over existing tractor tyres can increase ground contact area by a factor of 5   (Colmant Cuvelier)

DUAL TYRES

Dual tyres used without adding to tractor weight allow reduced tyre pressures and therefore larger ground contact area and reduced soil damage and field rutting (fig. 7.9).

At the same time they may give reduced efficiency because of higher rolling resistance. Where increased traction is required under comparatively firm conditions dual tyres will allow heavier axle loading. Then they may give increased pull of as much as 40 per cent compared with a single tyre at the same air pressure.

As with cage wheels, dual tyres need to be equal in size to be effective and they also need to be in a similar state of wear and inflated to the same pressure. They are best placed as close together as possible provided this does not trap stones and clods which may cause damage or initiate soil filling between the tyres. A scraper running between the tyres is often useful. In some cases wide spacing fittings are available to allow dual tyres to run between adjoining pairs of crop rows.

(Above) Wide spacing for row-crops

(Above) Detail of spacing adjustment for row-crops

(Left) Close arrangement for general field-work

Figure 7.9   Dual wheel arrangements   (F.A. Standen & Sons)

CHAPTER 8

TRAFFIC AND SOIL DAMAGE

Increasing tractor and implement weight and its effect on soil structure and crop growth has caused concern over most of the years of this century. The arrival of rubber tyres increased concern since ballasting of a basically heavy tractor became necessary to get traction. The problem of soil smearing by a slipping rubber tyre was recognised.

Adverse effects of traffic were noted long before the advent of tractors. Jethro Tull in the eighteenth century noted that people who overworked soil in a moist state made it like 'a highway', through frequent treading by horses. By the end of the nineteenth century subsoil tines attached to ploughs were used to break pans caused by horse hoofs and plough soles in the furrow bottom. The effect on crop yields of traffic at ordinary levels is difficult to show experimentally, although there are many well-documented case studies of severe effects of traffic on commercial farm crops where, possibly because of a difficult season or mismanagement, structure has been damaged.

The effect of traffic on the soil has been shown to be increased bulk density*, increased shear strength†, reduced porosity and reduced air and water permeability.

COMPACTION

Since the dry material in most soils is of fairly constant density

---

* Bulk density is defined as the weight per unit volume of dry soil and is usually expressed as kg/m³.
† Shear strength is defined in Chapter 7.

(about 2.65 times the weight of water), it follows that the density of a volume of soil in the field will increase or decrease according to the amount of air space that there is in the soil. Since the effect of traffic is to close up some of the larger air-filled spaces, measurement of the bulk density gives an indication of compaction. Weighting of the soil increases its strength and resistance to shearing. This is the reason for ballasting tractors for extra traction as explained in Chapter 7. Measurement of soil strength in the field is a good measure of compaction.

Porosity is a measure of the size and number of air spaces, and changes give an indication of structural damage (fig. 8.1). The large pores which are closed by traffic are also those which act as drainage channels through the soil. The rate of water intake is therefore a sensitive measure of damage. Gas permeability is also a sensitive measure but because of practical difficulties is rarely used to detect compaction.

MEASUREMENT OF PHYSICAL CONDITIONS

Density
Most methods of measuring density depend on weighing a known volume of soil. The known volume may be taken either by filling a cylinder of known size by pushing it into the soil and then taking out the soil inside it, or by refilling the hole from which soil has been removed with dry sand (of known density) or with water held in a membrane. The latest methods of density measurement depend on the measurement of radiation from a gamma ray source placed in the soil (photo 8.1).

Shear Strength
Laboratory methods of measuring shear strength are of limited value because the general aim is to measure the strength of the soil as it is undisturbed in the field. The apparatus consists of a crossed vane which is rotated in the soil, or a cylinder with internal vanes which is rotated in the same way (see fig. 8.2). The force required to shear the soil is measured with a torque wrench. The cylinder type equipment has the advantage that it can be weighted to different levels to give a measure of both cohesive and frictional soil strength. The vane type measures cohesion alone.

Penetration Resistance
The instrument used is the penetrometer which commonly takes

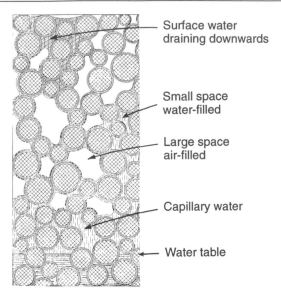

Surface water
draining downwards

Small space
water-filled

Large space
air-filled

Capillary water

Water table

*Figure 8.1   Diagrammatic representation of soil porosity. The large
spaces which do not hold water against gravity are those closed by the
compaction process*

Plate 8.1   Gamma ray equipment for soil density measurement
(Silsoe Research Institute)

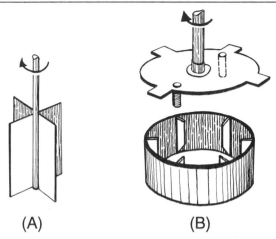

Figure 8.2   Measurement of soil strength: (a) shear vane; (b) shear box

the form of a metal rod about one metre long, graduated, and with a stainless steel conical tip. The upper end of the rod is fitted to a proving ring with a handle and dial gauge (photo 8.2). A proving ring is a standard steel ring which requires a known force to distort it a given amount. As the cone is pushed into the soil, the ring is forced out of shape, and the dial gauge gives a measure of the force. The graduations on the stem of the penetrometer give the depth at which the resistance was encountered. A constant rate of entry into the soil is required to obtain satisfactory readings, and in the more advanced instruments provision is made to assist the operator or automate this function.

Both shear strength and penetration resistance depend very much on soil moisture content and are generally of value as comparisons between different parts of the same field on the same day.

There are many forms of penetrometer including self-recording and vehicle-mounted models and of course the farmer's stick. Penetrometers and shear equipment are of little use in dry, hard or stony soils.

Water Infiltration

The measurement of water infiltration rate is a valuable check on compaction damage because it is sensitive and closely related to practical problems. Reduced infiltration rate leads to water saturation for longer periods during rain, with a damaging effect on plant growth and the performance of field machinery. The usual

Plate 8.2 The penetrometer for measuring soil penetration resistance; a simple model without a recording mechanism
(Silsoe Research Institute)

method employed is to place a graduated glass cylinder in the soil, fill it with water, and time the rate of lowering of the water level in the cylinder. This method is of more value where soils have been cultivated and are then being compacted following cultivation than on uncultivated soils where worm holes, root channels and various fissures can give unreliable readings.

Tractor Wheel Impression

The impression that a tractor driving wheel makes in the soil may be made up of soil that has been displaced around the tyre and soil that has been compacted beneath it. Under conditions of very high slip the rut may be partly re-filled as soil is moved from front to back of the tyre. Where the soil is not saturated and slip is not excessive, it can be taken that the depth of the wheel impression is a fairly accurate measure of the amount of compaction taking place.

TRAFFIC PATTERNS

The number of passes over a field by wheeled equipment during the production of one crop varies considerably but has been recorded at more than twenty times over for some horticultural crops and commonly involves wheeling 90 per cent of the field area (fig. 8.3).

Figure 8.3   Example of pattern of tractor wheel tracks during traditional seedbed preparation (area 3 × 3 m)   (D.B. Soane)

REPEATED COMPACTION

The first wheeling does most of the compaction (as much as 90 per cent of the total increase in density possible) (fig. 8.4) so that the greatest effect on the field is obtained from haphazard wheeling. This may not be a problem in many cases and on puffy soils may be an advantage. Where soils and crops are sensitive, however, a single wheeling may compact the soil to a level which affects the crop, and then some form of bed system may be valuable. Immediately after primary cultivation, the land is set out into beds of convenient tractor and implement width (the maximum practicable with axle extensions is likely to be 2.5 m) and is operated at those wheel centres for the rest of the season. No cropping will be

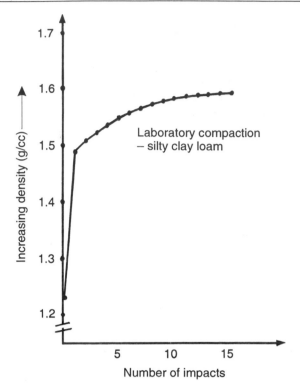

Figure 8.4   The first impact gives a large increase in density while following impacts have less effect

possible in the wheel tracks but the beds between will be completely free of wheel damage.

SOIL MOISTURE AND TRAFFIC

The effect of the amount of moisture in the soil on soil damage is greater than any effect likely to be encountered through differences in tractor weights, tyre sizes or ground pressures. Moisture content changes of the order of 2 or 3 per cent may have a profound effect on soil damage. Therefore strategic and day-to-day management of soil working operations is of primary importance, and is supported by optimum choice of equipment.

PLASTIC LIMITS: DEFINITIONS

The *lower plastic limit* of a soil is the minimum moisture content at which puddling is possible and the maximum moisture at which the soil is friable. It is usually defined as the moisture content at which the soil can be rolled into a 'worm' about 3 mm diameter without breaking.

The *upper plastic limit* is the moisture content at which the soil changes from a plastic solid to a viscous liquid. The test for the exact point is complicated and of little importance in farm practice.

MOISTURE AND COMPACTION

Dry soil is difficult to compact. As moisture content increases, soil becomes more easily compactible up to a certain optimum point. The actual position of the optimum point depends to some extent on the comparative force being applied. Above the optimum moisture content the compaction obtainable declines because the larger pores are partly water-filled and therefore less able to reduce in volume. The situation is shown in fig. 8.5.

It will be noted that the optimum moisture tends to be at or just below the lower plastic limit, which again is at or near the well-drained state (field capacity) in many soils and is near a moisture content generally considered favourable for ploughing. Where the plastic limit is below field-capacity moisture content, soils are particularly difficult to manage. Soils where the plastic limit is above field capacity are considerably less difficult.

PUDDLING AND SMEARING

Although soil may not compact further through straight applied pressure once the optimum is reached the strength of the soil continues to decline as moisture increases further. Then wheel slip tends to increase with attendant puddling* and smearing†, and the bearing strength of the soil declines to give increased sinkage, with

---

* Puddling—the mechanical process whereby aggregates of wet soil are disrupted and some clay is dispersed.
† Smearing—localised spreading and smoothing of a soil by sliding pressure.

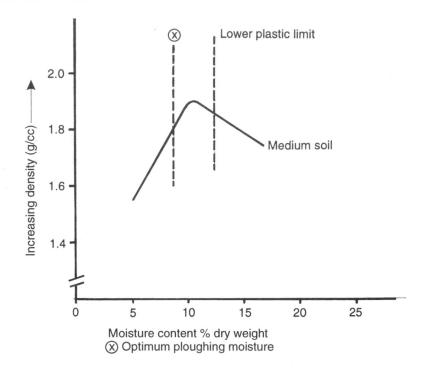

Figure 8.5   The optimum ploughing moisture and lower plastic limit is shown in relation to the moisture content giving greatest compaction

increased rolling resistance and again greater puddling and smearing. The result is inefficient operation and the likelihood of severe soil damage.

The obvious aim must be to keep off soil that is wet and that will be difficult or impossible in many years and cropping situations.

Where it is necessary to have traffic on wet soil it should be noted that:

- soils that have been newly cultivated are mechanically weak and therefore more liable to damage than a settled soil (such as a stubble) at the same moisture content;
- well-structured and freely draining soils are at risk for shorter periods after rain than those that are slow draining, either naturally or because of previous compaction. Crawlers have a marked advantage in reduced soil damage compared with wheeled tractors of whatever weight or tyre size.

TRACTOR WEIGHT, TYRE SIZE AND WHEEL SLIP

It was indicated in Chapter 7 that the stiffness of the carcass of a tractor tyre carried only a small proportion of the weight of the tractor. The remainder is in effect borne on the air pressure inside the tyre. Therefore, if a given air pressure is employed and tyres are sized so that varying weights of tractor can be carried, ground pressure would be similar for a range of tractors. This situation does not hold entirely true because some of the larger tractors run at high inflation pressures in order to carry the tractor on tyres of moderate size for ploughing in the furrow. A wide range of tractors do, however, apply very similar ground pressures.

If the ground pressure is constant, the effect on the soil will depend primarily on moisture content and porosity (in effect looseness) of the soil, and only secondarily on the size and shape of the contact area.

PRESSURE PATTERNS

The pressure effects of tyres on the soil have been calculated and verified experimentally. Fig. 8.6 shows examples of the lines of equal pressure which give the characteristic patterns.

In this discussion no account is taken of the local variation in ground pressure beneath tyre lugs. In very wet soil pressure beneath the lugs may not differ greatly from the average pressure under the tyre. As the soil becomes harder and drier, the pressure under the lugs increases to become 3–4 times the average pressure under very hard soil conditions. Under very hard conditions very high pressures will not cause compaction because the soil obviously has sufficient strength to withstand these pressures. Pressure tends to be higher towards the edge of the tyre due to the stiffness of the wall, and this is particularly noticeable at very low tyre pressures.

REGION OF STRESS

The facts emerging from the stress pattern beneath the tractor tyre are:

- the stress tends to concentrate under the centre line of the tyre (the load axis) and the tendency is greater with increased

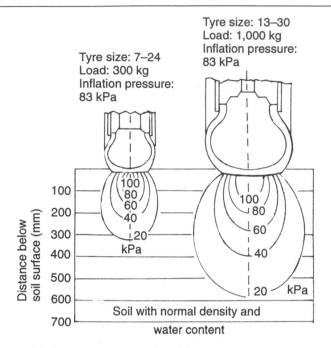

Figure 8.6 Calculated curves of equal pressure under tractor tyres; an example of increased load at the same tyre pressure (ASAE)

moisture and reduced cohesion. That is, the tendency becomes greater as the soil becomes weaker.

- The region of maximum stress is not immediately at the tyre face but is some distance below the soil surface. The actual depth at which this maximum stress occurs increases chiefly with increasing moisture content, but also with surface pressure and total wheel load.

While ground pressure is important for the maximum level of compaction involved, the depth to which compaction takes place depends on total wheel load (fig. 8.6).

For example, if the soil under tractor wheelings during spring work is examined it will be found that where a light tractor with a given tyre pressure compacts soil say 100 mm deep, a heavy tractor at the same tyre pressure will compact it fully to ploughing depth. This is why tyre sizing cannot fully eliminate load effects and why, for extreme conditions, vehicles combining low weight with low ground pressure are required.

1   Symptoms of acidity in oilseed rape. Marginal yellowing and cupping of leaves is typical of this condition in all brassicas

2   Manganese deficiency symptoms in oats. Light rusty lesions coalescing into scorch-like symptoms known as Grey Speck

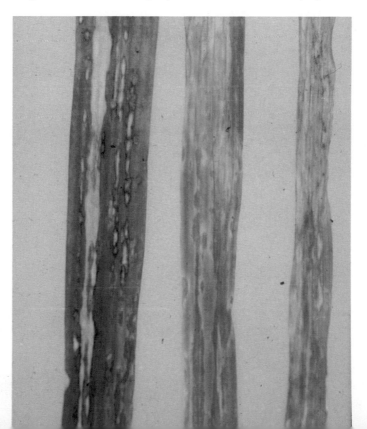

3 Field of spring barley with overall manganese deficiency except in wheelings where consolidation has controlled the symptoms

4 A sugar beet plant with boron deficiency. Yellowing of younger leaves and death of growing point

5   Growing point of oilseed rape showing sulphur deficiency.
Interveinal yellowing of younger leaves

6   Potato leaf with symptoms of magnesium deficiency.
Interveinal yellowing with some dark brown spotting

7   Potato crop showing magnesium deficiency-like symptoms. Roots restricted by a severe pan at 30 cm in a heavy soil well supplied with indigenous magnesium

8   Severe rill erosion in a potato field caused by rain falling on wheelings down a 10 degree slope

9   Sugar beet growing on peat protected from wind erosion by rows of mustard, shortly before destruction

10   Wind erosion control on a sand at ADAS Gleadthorpe. Sugar beet drills are seen at right-angles to the small ridges produced by a furrow press used at ploughing

11   Very large 'horsehead' clods after ploughing a silty clay in a dry summer. Indifferent farming for several years led to this poor structure

12   Good structure in the topsoil of a chalky boulder clay ideal for shallow cultivation. Note good rooting, many cracks, rough surfaces and warm brown colour

13   Poor structure in the topsoil of an Oxford clay in need of loosening. Note the paucity of roots, smooth dense faces and grey-brown colour

◀ 14 The best soil structure the authors have ever seen! Ideal friable conditions to 1 m on the slopes of a Lake District hill, with an acid humus surface

15 Poor conditions in a Hanslope clay profile. Note the dense areas in the topsoil and the 20 cm deep pan immediately below the topsoil ▶

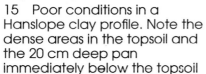

◀ 16 Plough pan in a medium textured soil. Note the shiny furrow bottom caused by worn plough shares and slipping tractor wheel

17 An anaerobic pan at the base of the plough layer caused by ploughing after lifting sugar beet. Note the grey colour caused by reducing conditions

18 Cereal land waterlogged over a pan, with ditches nearly empty and drains not running

LOW GROUND PRESSURE VEHICLES

Developments in pesticides have created a demand for vehicles which will traverse saturated soils and growing crops with minimal soil and crop damage. The requirement is for ground pressures as low as 35 kPa, for a vehicle weighing probably 1 tonne or less, with a carrying capacity up to 120 per cent of its own weight, a form of steering that does little or no damage on the turns, adjustable track and clearance up to 1.0 m. The high clearance is desirable so that the vehicle can be used for chemical and fertiliser application later in the growing season when clearance rather than low ground pressure is important.

A wide range of vehicles, each offering some or all of the requirements suggested above, is available. 'All Terrain Vehicles', three- or four-wheel lightweight tractors based on motor-cycle engineering, fulfil the basic requirement for many farmers. In many cases alternative tyre equipment for existing tractors or cross-country vehicles provides an adequate solution to the problem. Low tyre inflation pressures are an important factor in achieving low ground pressures and claims made for low ground pressures from comparatively high-pressure tyres should generally be queried. Exact measurement of ground pressure under tyres is difficult, and the pressure under any one tyre will vary widely with soil conditions (fig. 8.7). Long, narrow ground contact areas, as in

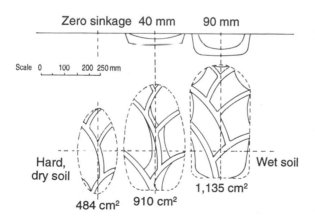

Figure 8.7   Contact areas for one tyre under different soil conditions. Tyre pressure 80 kPa, load 700 kg   (ASAE)

multi-wheel vehicles and crawlers, are more effective than the short and wide pattern of single large wheels.

EFFECT OF PRESSURE

The actual ground pressure at which compaction affecting plant growth is likely to be caused cannot be specified. The range of pressures that the soil will stand varies considerably, and even for soils that one would consider to be in the usual working moisture range, there are records of pressures as low as 30 kPa causing trouble, and 240 kPa not causing damage. It has been previously mentioned that newly cultivated land is particularly vulnerable, and there have been cases of newly ploughed land collapsing under very low applied pressures.

The vibration effect of traffic has been shown to have a compacting effect, but it is of little practical importance and only applies to some sandy soils.

The effect on the crop of the soil damage and reduction of size and number of larger pores is seen in two principal ways. Water movement is slow or negligible in compacted soil, and the roots of crops are restricted by anaerobic (oxygen-less) conditions associated with temporary waterlogging. The damage to the roots is probably caused by gases such as ethylene and hydrogen sulphide produced by decomposition under these conditions rather than by a direct lack of oxygen. The second effect on the roots is the result of the reduction in large pores and the increased mechanical strength of compacted soil. It has been shown that in an excessively strong soil roots can only grow through pores of greater diameter than the root tip. Therefore, if the soil is strong but not compacted, as perhaps in drying clay, roots can grow. It has also been shown that lack of the larger pores is not a problem if the strength of the soil is not too great. This was demonstrated by growing cotton roots through a 25 mm thick layer of wax, which had no pores but was not mechanically strong, and it was concluded that mechanical strength was more of a problem in compacted soils than the lack of large pores.

HARVEST TRAFFIC

Late autumn root harvesting often involves very wet conditions with heavy trailers fitted with high-pressure tyres. There is generally very little that can be done to alter the date of harvest or the type of equipment, and the best recommendation is to apply remedial measures to the soil as soon as possible after such a harvest.

MINIMISING SOIL DAMAGE

The most the farmer can do to limit compaction damage on vulnerable soils is:

- avoid moist and loose soils as far as possible;
- cut down traffic to the minimum necessary to grow the crop;
- ballast tractors to reduce wheel slip for operations involving heavy draught;
- remove ballast from tractors when it is not required;
- use crawler tractors whenever practicable;
- use dual tyres, extra large low-pressure tyres (photo 8.3), cage wheels, half-tracks or special-purpose vehicles;
- in extreme cases resort to bed systems of cultivation.
- use higher speeds and lower draughts whenever practicable.

Plate 8.3   Massive 66 x 43.00 low pressure, low-profile tyres to minimise pressure applied to the soil   (Arable Farming)

CHAPTER 9

PLOUGHS AND PLOUGHING

The mouldboard plough has for many years been the basic tillage tool in Britain, and its chief asset is the ability to invert the soil, so covering trash and weeds and exposing fresh soil at the surface. Further advantages are that the entire area of the field is cultivated, whereas alternative implements may leave uncultivated or partly cultivated sections between tines, and land can be 'set up' to assist drainage, a wide range of degree of pulverisation of the soil can be achieved and, from the operator's point of view, ploughing is rewarding for the highly skilled workman.

The basic form of the plough is a share to make the horizontal cut through the soil, the coulter to make the vertical cut, the mouldboard for turning the furrow, and the landside to take the sideways thrust as the furrow is turned (fig. 9.1). It is over 200 years since the iron mouldboard was first introduced. Early in the nineteenth century the chilled cast-iron self-sharpening share was invented, and soon afterwards ploughs that could be taken down into their component parts by the ploughman were introduced. Until the end of the nineteenth century ploughs were of the general-purpose pattern. Then the digging body, aimed at maximum soil pulverisation, was introduced and today's basic plough patterns were established.

The mouldboard plough is highly efficient as a means of moving a given volume of soil and may require as little as one-sixth or one-eighth of the energy that a rotary cultivator would require to move the same volume. The chief reason for this is the very limited amount of energy that is used for accelerating soil particles by the plough—at low speeds as little as 4 per cent compared with about 30 per cent of total energy input for the rotary cultivator.

Further information on power requirement and the effect on tilth produced is set out in Chapter 12.

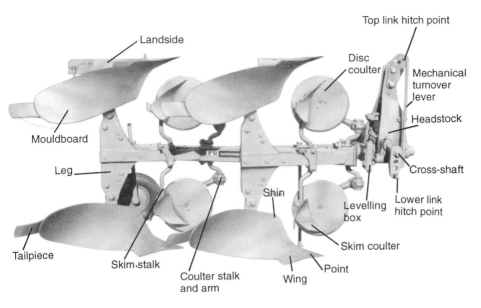

Figure 9.1   Basic plough parts   (Agrolux-Ransomes)

THE FUNCTIONS OF A PLOUGH'S SOIL-ENGAGING PARTS

Shares

Plough shares may be of steel or chilled cast iron and their function is to undercut and commence the raising of the furrow slice. The angle of entry to the soil is very low, making for low draught. Soil slides smoothly over the upper face giving low friction between soil and metal, and the lower side is shaped to avoid smearing of the furrow bottom. Severely worn shares with a rounded leading edge lose their ability to penetrate hard soil and are more likely to smear the furrow bottom. Further, rather than allowing smooth sliding of soil over metal, they tend to build up a cone of soil which increases draught and may lead to a low-pressure area further back, which in sticky soils will cause poor scouring of the mouldboard.*

The share exerts an effect on the plough and ploughing out of all proportion to its physical size, and should always be in good order. Replacement is generally the most economic remedy except for the large steel shares, which may be worth facing or rebuilding

* See page 135.

with hard weld. Shares may be formed in one unit or in three sections known as the point (itself sometimes called the share), the wing and the shin.

The Central Area of the Mouldboard

The furrow undercut by the share is lifted on to the central area of the mouldboard where, if the soil is friable, it is broken up. In clay soils the curve needs to be long and gentle since the soil is resistant to shear and, when moist, tends to form planes of weakness in the furrow slice rather than complete breakage. In light soils the curve should be steeper since the furrow collapses easily and a sharp curve is needed to keep it in contact with the mouldboard prior to inversion. The effect this area of the mouldboard has depends very much on speed of operation. General-purpose bodies, with a gentle curve, produce intact furrows at low speeds and broken work of the kind normally produced by the semi-digger body at higher speeds. The degree of pulverisation produced by other plough bodies also increases, with an increase in draught, at higher operating speeds (see fig. 9.7 and photo 9.1).

Plate 9.1   Effect of increased speed on work of TCN plough body   (Agrolux-Ransomes)

The Inversion Area of the Mouldboard

The rear end of the mouldboard is responsible for turning the furrow over. If it has too much turn in it for its normal operating speed the soil is thrown too far and unnecessarily high draught is involved. If it has too little turn a satisfactory inversion will not be obtained unless the operating speed is high.

TYPES OF MOULDBOARD

General Purpose, Helicoidal or English Type
This is long, with a gentle convex curve and is not intended to pulverise the soil. It is particularly suitable for ploughing out grassland. At the low forward speeds for which it is intended, it produces 'set up' furrows which under some conditions can be an aid to surface drainage (photo 9.2). It is not intended for deep working, the maximum depth considered acceptable being two-thirds of furrow width. When operated at increased speed the side throw of soil is considerable and draught and degree of soil pulverisation increases sharply.

Plate 9.2 Ploughing using a general purpose mouldboard
(Agrolux-Ransomes)

Digger, Cylindrical or Continental Type
This is short, abrupt and concave, and is shaped as part of a cylinder. The abrupt curve compresses and crumbles lighter soils

and tends to break heavy soils into large clods. The furrow can be at least as wide as it is deep, and inversion is good. This mouldboard is not suitable for ploughing out grassland. High-speed mouldboards have been developed from this basic type.

A whole range of mouldboard forms lie between these two extremes, the best known being the semi-digger. This is short, deep and is slightly concave in shape. The furrow slice is more broken than the general purpose type and it will invert a proportionately deeper furrow (photo 9.3).

Plate 9.3   Ploughing using a semi-digger mouldboard
(Agrolux-Ransomes)

PLOUGH FRAME DESIGN

Trailed Ploughs

Ploughs may be trailed, semi-mounted or mounted. The trailed plough is rarely seen. As the name implies, it uses the tractor solely for the provision of draught. All adjustments for depth and level of ploughing are contained within the implement. They do not require 3-point linkage equipment on the tractor and can be built to very large sizes without suffering a weight limit by the capacity of the tractor hydraulic system. They are therefore well suited to many of the larger and older crawler tractors. The cost of a trailed plough is of the order of 30 per cent more than a mounted plough of similar capacity and it suffers a considerable disadvantage in manoeuvrability and ease of transport.

Mounted and Semi-Mounted Ploughs

Mounted and semi-mounted ploughs make use of the tractor linkage equipment. The fully mounted plough is comparatively inexpensive, highly manoeuvrable, accurately controlled by tractor hydraulic systems, and its own weight can give weight transfer and therefore better traction from the tractor.

The weight of a mounted plough tends to pivot the tractor around its rear axle. If the plough is large and the front end of the tractor light, the front of the tractor will leave the ground. This can be balanced up to a certain point by increased weighting of the front of the tractor (fig. 9.2). As the plough becomes larger not only does its weight increase, but because the centre of gravity moves further away from the axle, the effect of that weight multiplies up very quickly. Fig. 9.3 shows the very rapid increase in the implement moment or levering effect with each additional mouldboard.

While tractors have rapidly increased in engine size the stability moments or resistance to the levering effect have not increased at the same rate. There are therefore cases of tractors being able to

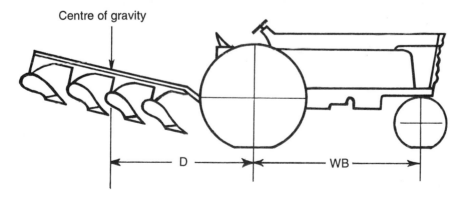

Figure 9.2   Stability depends on the front-end tractor weight ×
wheelbase (WB) being greater than plough weight × D   (ASAE)

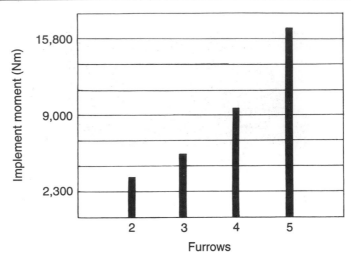

Figure 9.3   Showing the rapid increase in levering effect on the tractor of extra furrows on the plough   (ASAE)

pull more furrows than they can lift or than their stability will allow if fully mounted.

Semi-mounted ploughs overcome this problem by mounting the rear of the plough on a wheel while using the lower or all three links on the tractor. This type of plough has been particularly provided for in electronic linkage developments, and allows hydraulic control of ploughs to match the largest wheel tractor. A further advantage is in its ability to follow more pronounced curves than large full-mounted ploughs, but this applies chiefly to countries where contour farming is practised.

Clearances

The total length of the plough is dependent on the front-to-rear spacing of the mouldboards, which in turn depends on the method of making the vertical cut for the furrow slice. Adequate clearance between mouldboards is necessary to allow room for trash to fall into the previous furrow bottom before being covered by the following furrow, and to ensure that extra draught is not incurred by pressing the following furrow against the landside of the previous plough body (fig. 9.4). The usual range is 750–1000 mm. Clearance under the frame of the plough is important for avoiding clogging with trash. The clearance normally provided under British conditions is 650–750 mm, which successfully deals with chopped straw and all other crop residues.

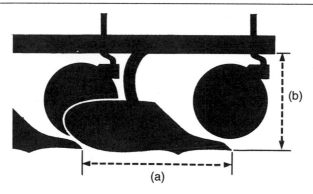

Figure 9.4   Plough clearance: (a) between bodies; (b) under the beam
(Agrolux-Ransomes)

Coulters

The vertical cut for the furrow slice is normally made by a coulter attached to the plough beam. It may be a simple knife arrangement or a plain, scalloped, or wavy-edged disc (fig. 9.5). All forms of coulter are fully adjustable for position in relation to the plough share. Knife coulters are simpler in construction than discs and allow the design of a shorter plough with less strain on the tractor hydraulics. They are more likely to cause blockages under conditions where there is much surface rubbish and disc coulters are more generally used in Britain. Wavy-edged and scalloped discs are an advantage where trash blockages are a problem.

Coulters are not universally used. Under some conditions a knife coulter is used on the last furrow only; the mouldboards themselves may be specially shaped to give a vertical cutting action or there may be no form of coulter. In each of these cases there will be some advantage in being able to shorten the plough and bring its centre of gravity closer to the tractor.

Skim Coulters

More effective burial of rubbish can be achieved by slicing off the shoulder of the furrow immediately before it is turned and dropping it into the previous furrow bottom. This is done with skimmers, or skim coulters (photo 9.4), a miniature form of mouldboard which may be directly mounted on the plough frame or as part of the coulter assembly. The former system is more satisfactory where there is an appreciable amount of straw or manure to be ploughed in and where high-speed ploughing is contemplated.

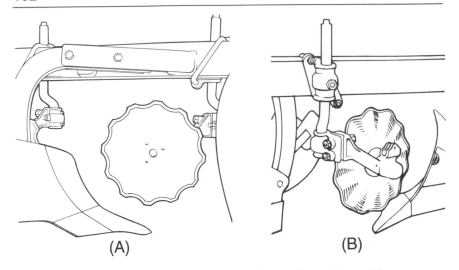

(A)   (B)

Figure 9.5   Alternative forms of coulter: (a) scalloped disc; (b) wavy edged disc   (Agrolux-Ransomes)

Plate 9.4   The action of skim coulters   (Agrolux-Ransomes)

Skim coulters are not favoured very often by ploughmen and spend as much time in the shed as on the plough. This is probably because, through incorrect use, they either cause trouble or are of no apparent advantage or because as ploughing speeds increase they are the first element of the plough to cause blockages. The main points of skim setting are that they should pare off the minimum depth of soil necessary to take off the rubbish, and they should be set far enough forward to allow the material pared off to fall to the furrow bottom before the furrow is turned by the mouldboard.

Adjustments

Provision is made in the plough for adjustment of the front furrow width and levelling in both directions. It is often possible to change the furrow widths by alternative bolting positions for the legs supporting the plough bodies. Some ploughs offer on-the-move adjustment of furrow width setting, or pre-setting of the width of right and left furrows, which is a particular advantage on hilly land. Furrow width is reduced for uphill work, and extended when working downhill (photo 9.5).

Plate 9.5   Semi-mounted plough – the Vari-width   (Kverneland)

Safety Devices

Where damage may be caused by obstructions, soft shear bolts can be fitted so that plough bodies may 'break-back' on impact. These have proved less satisfactory than spring-loaded break-back devices which are not only more sensitive but are easier to re-set than shear bolts. It is often an advantage to be able quickly to reduce or increase plough size by a furrow. Some ploughs offer this facility, although making the change may not be particularly quick or easy.

Reversible Ploughs

Most ploughs today are reversible—that is, they are designed to turn the soil to either right or left. The advantages they give are that they leave a level field after ploughing and there is less skill required for setting out the land. They do not generally give any advantage in work rate, they are expensive since they carry a second set of bodies, and are heavy so that fully matching a large tractor with a mounted reversible plough may not be possible, and semi-mounting will be necessary. Semi-mounted reversible ploughs of up to ten furrows are now available. Turn-over mechanisms are normally hydraulic, but may be mechanical.

Offset Linkage

Wheeled tractors normally plough with the work-side wheels in the open furrow. The greatest advantages of this system are having the line of draught of the most common plough sizes central to the tractor linkage and having, through the 'feel' of the furrow wall, assisted steering. Difficulties arise where high-power tractors are fitted with tyres considerably wider than the furrow. The largest wheel tractors plough with all wheels on unploughed land, with ploughs large enough to give a satisfactory line of draught and no excessive lateral pull or 'crabbing'.

Concern for the effect the tractor tyre may have on the open furrow bottom has encouraged developments in ploughing on the land at various times. The chief technical problem has been tractor steering and the tendency of the furrow wall to collapse under the pressure of the nearby tyre. Automatic steering systems, sensing the furrow wall, have been developed and the crabbing effect of the plough has been countered by a large disc fitted at the rear of the plough. Mounted ploughs which can be offset on the hydraulic linkage for on the land operation are commercially available (fig. 9.6 and photo 9.6).

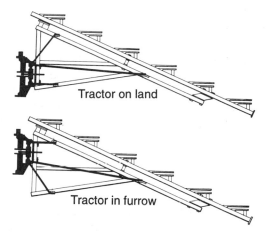

Tractor on land

Tractor in furrow

Figure 9.6   Offsetting arrangement for ploughing on the land
(Agrolux-Ransomes)

Soil Conditions and Ploughing

The mouldboard plough is used as a primary cultivation implement under practically all conditions of soil moisture and texture.
Under very dry conditions in heavy soils, penetration may be a
problem and wear on shares and mouldboards may be high.
Subsoiling has in some cases been resorted to in order to make
ploughing possible. Ideal ploughing conditions are when the soil
is moist but still below the lower plastic limit. Heavy soils do not
respond to pulverisation by the more concave type mouldboards
and higher ploughing speeds, and it is only on lighter soils that
seedbed conditions for cereals can be achieved by ploughing
alone.

The chief ploughing problem caused by particular soil conditions is a build-up of soil on the mouldboard, or failure to scour.
This results from the physical strength of the soil not being great
enough to overcome its own adhesion to the mouldboard and is
most likely to occur on poorly structured silty clay loams. The
problem does not occur in dry soils, which do not tend to stick and
are physically comparatively strong. Very wet soils are unlikely to
stick because there is enough water present to act as a lubricant
between soil and metal.

General-purpose bodies are more likely to scour clean than the
more concave types because of their smaller angle to the direction of
travel. Some advantage has been gained from removing the coulters to increase the cleaning pressure that the soil exerts on the

Plate 9.6   Adjustable headstock working in the in-furrow
position   (Agrolux-Ransomes)

mouldboard. While differences in the metal used in the mouldboard
have no appreciable effect, some advantage has been gained from
the use of low-friction plastic coverings, and reduction in contact
area through slatted or specially small mouldboards.

MATCHING PLOUGHS AND TRACTORS

Tractor power can be utilised for ploughing in terms either of
more furrows at low speeds or fewer furrows at high speeds.
Crawler tractors are able to develop high horsepower at low

speeds and, provided a large enough plough is available, can operate efficiently in this way. By contrast, wheeled tractors are able to develop their full power more efficiently at higher speeds and lower draughts and are therefore likely to be more effective with fewer furrows. There is, however, a factor which goes against this efficiency and that is the increase in plough draught generally associated with increased forward speed.

SPEED OF PLOUGHING

Experiments on the subject have been carried out over most of the years of this century (fig. 9.7). During the first quarter of the century the average increase in draught was found to be about 20 per cent for an increase in speed from 3.2 to 6.4 km/hr. In more recent years the average calculated from several hundred trials at 1.6–9.6 km/hr was 25 per cent between 3.2 and 6.4 km/hr and 50 per cent between 4.8 and 9.6 km/hr. Ploughs tested in Britain in recent years have shown slightly lower figures than these—of the order of a 20 per cent increase in draught for a speed increase from

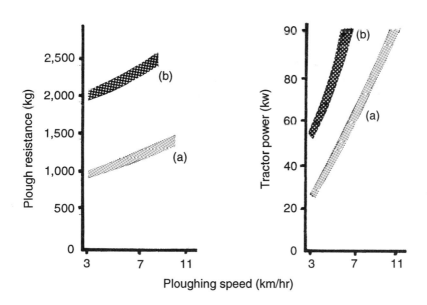

Figure 9.7   Draught and tractor size required by four-furrow plough: Light soil (a) 0.35 kg/cm². Heavy soil (b) 0.75 kg/cm²   (ASAE)

4.8 to 6.4 and from 8.0 to 9.6 km/hr. In these records there are cases of ploughs showing only about 5 per cent draught increase where one may have expected 20 per cent. These were in the heavier soils where the higher speed did not give any extra pulverisation of the soil.

The extra power is taken up chiefly in accelerating soil particles to no useful effect, the force required increasing as the square of the speed. Thus twice the speed gives four times the acceleration force. The force required for cutting the furrow slice does not vary appreciably with speed. Not all the extra power is lost. The extra pulverisation achieved may save later cultivations.

The work rates achieved at public demonstrations often show the folly of attempting to pull too wide a plough. Wheeled tractors are under many conditions best operated at about 7 km/hr for ploughing, and it may be necessary to cut down plough size to achieve this speed. It may be worth pointing out that in some trials of existing plough bodies at high speeds, the upper acceptable speed limit has been determined by excessive pulverisation rather than increased draught. Generally ploughs that are capable of ploughing at 8–10 km/hr have to be operated at high speed in order to do a satisfactory job.

SYSTEMS OF PLOUGHING

One-way Ploughing

The reversible plough allows the simplest possible system of ploughing. A start is made at one side of the field and ploughing progresses to the far side. The plough is reversed at each end of each furrow. Some adjustment to the plough may be necessary in order to get even matching of adjoining sets of furrows, but once this is done the field can be ploughed out without any ridges or open furrows to break the evenness of the surface (fig. 9.8).

The generally recommended method of turning is the reverse turn (fig. 9.9). It may be quicker, however, to leave a headland wide enough to make a loop turn. The difference in turning time quoted for these two methods is of the order of 0.3 minute for the loop against 0.6 for the reverse turn. The general rules are as follows.

Turning time depends largely on forward speed and the turning circle of the outfit. For this purpose a fully mounted plough can be considered part of and having the same turning circle as the tractor.

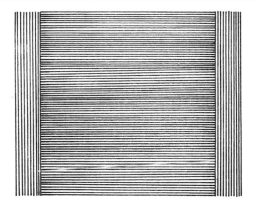

Figure 9.8 The simplest form of one-way ploughing

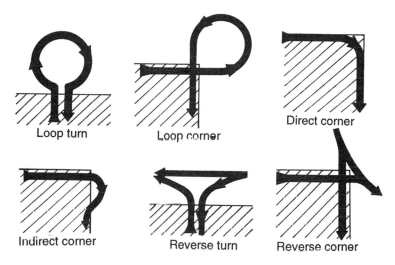

Figure 9.9 Methods of turning applying to most field operations.
(M. Upton)

A loop turn and a loop corner take the same time. A reverse turn is slowest and a direct corner fastest of all.

It is worth noting that setting out a sufficiently wide headland is likely to make for faster turns at the furrow ends and will not slow down the overall work rate, because most headland ploughing involves direct corners and therefore does not require reversing to get round corners.

Ploughing in Lands

Orthodox ploughs turn furrows in one direction only, and the traditional method of operation aimed at keeping the 'empty' running time to a minimum is known as ploughing in lands (fig. 9.10). The field is set out in parallel strips by ploughing up ridges at equal distances apart. These are then ploughed around— the process being known as gathering—and at a later stage the unploughed section may be ploughed around, when the process is known as casting. Either casting or gathering too large an area gives excessive wasted time in headland running, and a system is in use which aims to combine casting and gathering in the most economical way. This depends on operating on each land in four equal sections and keeps headland running down to a maximum of half a land width. The land width is usually 30–50 metres. However, it may be up to 100 metres depending on the size of the plough.

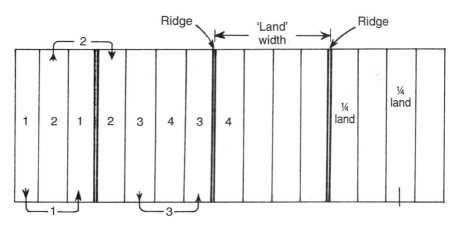

Figure 9.10   Ploughing in lands. The ridges are set up parallel and equal distances apart. The 'quarter lands' in the figure are numbered to indicate the system adopted to keep idle running time to a minimum   (Agrolux-Ransomes)

Where fields are irregular in shape the ridges are set up parallel to the longest and straightest side of the field. The only snag with ploughing in lands is the unevenness of the ridge at the start and the furrow at the finish of each land. It is customary to make the new ridge at the site of the previous year's furrow in order to minimise this problem.

Round-and-round Ploughing

Considerable reduction in the use of ridges and furrows can be made by round-and-round ploughing. This may be done by marking a plot in the centre of the field, with boundaries equal distances from the field boundaries. This is then ploughed around a central ridge and forms an island around which the whole field is ploughed. The marking out may be difficult and a simpler system is to plough from the outside of the field inwards. The systems are illustrated in fig. 9.11.

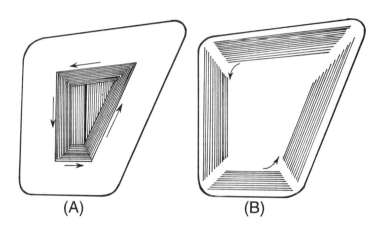

Figure 9.11   Round-and-round ploughing: (a) starting from the centre; (b) starting from the outside. The angled corner strips are finally ploughed out

Disc Ploughs

The term plough has been used throughout to indicate a mould-board plough. In some countries disc ploughs are important (photo 9.7). They consist of a series of discs generally up to 900 mm diameter which are angled to the direction of travel and some-times also angled to the vertical. Draught is about 10 per cent higher than the mouldboard for the same work; they do not give the same uniformity of depth of work or finish and do not entirely bury trash in the manner of the mouldboard plough. They have the advantage of greater resistance to damage from obstructions (which they tend to ride over) and generally give better penetra-tion in very hard soils. (See also page 151.)

Plate 9.7   Detail of a disc plough. The disc is angled to the vertical and the line of travel   (Parmiter)

Swing Beam Ploughs

Several manufacturers offer simple and robust reversible ploughs which utilise double-ended mouldboards on which the direction of soil-flow is reversed by hydraulically swinging the plough beam at the end of each furrow. In one case the angle of the plough bodies is additionally altered through a parallel linkage arrangement as the beam angle is changed. These ploughs provide the advantages of the one-way plough without the complication of a second set of bodies. They are lighter, and can be expected to produce less wear and tear on the tractor linkage; they have a lower centre of gravity and should be safer; and they are very simple to set up. They do not produce exactly the same quality of work as the traditional plough, and are probably complementary rather than a direct competitor.

CHAPTER 10

CULTIVATION IMPLEMENTS: Basic Forms

The main factors affecting choice of cultivation equipment are the type of tilth required, soil type and the rate at which the tilth has to be produced. Power available may affect the size but rarely the type of implement chosen. Since there are no scientific standards in use to describe quality of tilth and since the type of work produced by any one implement varies considerably in a particular soil with moisture content and density, the decision on which implement to use depends on the personal judgement of those in charge. An understanding of the principles involved may, however, be an aid to that judgement. Tined cultivation equipment can best be classified according to the angle of approach of the tine to the soil. A number of possible positions are shown in fig. 10.1.

An acute angle of approach is illustrated at (a). At this angle, and at angles up to 45°, the tine will tend to lift the soil and therefore part the particles by pushing upwards. Compaction is unlikely to occur and the lifting of clods or subsoil to the surface can in some cases be a problem. The draught at angles (a) and (b) is considerably lower than that of tines above about 50°.

Illustrated at (b) is the angle at which there is no marked tendency either to lift or exert a force downwards on the soil. While there is a substantial draught advantage over greater rake angles, this one is already on the steeply rising section of the graph (fig. 10.2), and for practical purposes is probably just over the best compromise angle. (c) illustrates the nearly upright situation where there is no tendency to bring material from below to the surface but where the downward force is not great enough to be a serious compaction risk. Represented at (d) and (e) are backward-sloping tines whose chief characteristic is their considerable down-

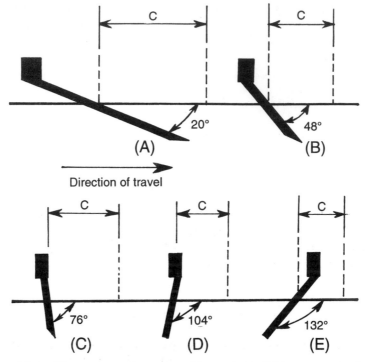

Direction of travel

Figure 10.1   Tine approach angles. The extent of the crescent of soil disturbed is indicated by 'C'   (Payne 1956)

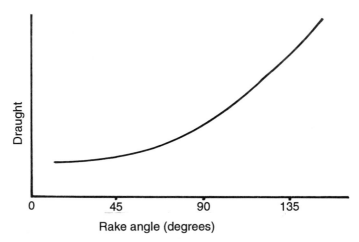

Figure 10.2   The effect of increasing rake angle on implement draught   (Payne 1956)

ward force on the soil. Whereas in (a) the cultivation effect of the tine was in lifting and thereby parting particles, the cultivation effect with backward sloping tines is crushing or shearing due to compressive force.

PATTERN OF SOIL CLEAVAGE

It has been shown that the pattern of soil cleavage is similar for the range of rake or approach angles quoted. A crescent of soil is disturbed in front of the tine, the length of the crescent being greater at the shallower approach angles but always closely related to the position of the point of the tine. A triangular wedge of soil is seen within the crescent up to a 100° angle, but disappears when the tine becomes markedly backward sloping.

These general observations hold good for a wide range of widths of tine and depths of operation.

WIDTH OF SOIL DISTURBANCE

The width of the area disturbed by a tine does not increase in proportion to depth of operation. This is shown in fig. 10.3.

Increases in depth of operation give large draught increases, however, and in practice this means that it is important in all

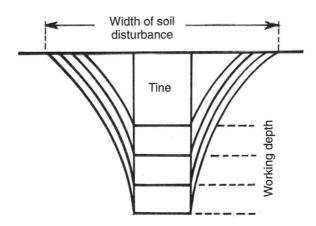

Figure 10.3   An increase in depth of work does not give a corresponding increase in disturbed soil width   (O'Callahan 1964)

cultivation to ascertain how deep the work needs to be and not to work beyond that depth. Where shallow overall cultivation is required in one pass, it is best achieved by closing up tine spacing or by the use of broad A-shaped cultivator shares. Another alternative is to use discs (page 151).

TINE SHAPES

Tine shapes available to farmers fall into three main groups, as shown in fig. 10.4.

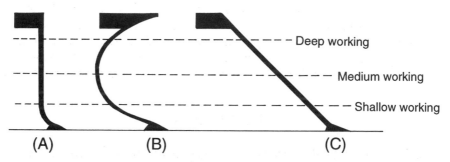

Figure 10.4 In shallow work the three tines approach at a similar angle. The effective angle of (a) and (b) increases with depth of work (G. Spoor)

In considering the many tine slope angles available it should be noted that the general principles set out above have to be considered in the context of the working depth of any one implement.

In fig. 10.4 the angle of approach in shallow cultivation may be effectively 45° for (b) and (c). In deep cultivation only (c) will be 45°. Both (a) and (b) may perform as upright tines, and if draught is high enough (b) may spring back to become greater than 90°.

HEAVY CULTIVATORS OR CHISEL PLOUGHS

A series of outline shapes of tines available to farmers are shown in fig. 10.5.

These are sold spaced on the implement frame at intervals of from just under to just over 300 mm. Since under many conditions chisel ploughing at 300 mm spacing does not disturb the entire

Chisel or heavy cultivator tines

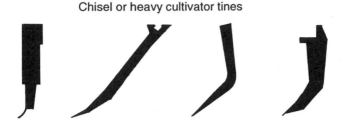

Figure 10.5   Outlines of some tine shapes in commercial use

field area, there is a case for having enough tines available to operate closer than 300 mm when necessary. It may still be necessary to cross-work to get full soil disturbance even at 250 mm spacing, and if that is so, the work is best done 'on the skew' for the second run rather than at right angles to the original work.

PENETRATION

Penetration is best with shallow approach angle tines provided they are rigid or, if spring loaded, the springing is strong and preferably adjustable. Where sprung tines cannot be tensioned sufficiently to hold their correct angle in very hard soil, they can be very poor penetrators. While penetration characteristics are important in heavy soils in dry seasons, good depth control, either from the tractor or by depth wheels, is equally vital. If heavy cultivators or chisel ploughs penetrate too deeply at the first pass, they may pull up clods that are difficult to break down afterwards.

FRICTION

The quality of the metal in cultivator tines is important from the point of view of length of life but not for draught reduction through reduction of friction. The various grades of steel in use, once polished by the soil, will give very similar results.

It will be seen from fig. 10.5 that the tines commercially used represent all sorts of combinations of approach angle, often in different sections of one tine. The likely performance of each one can then be judged from a consideration of the basic principle involved and the assumed depth of operation. The angle of approach of the point or share on the tine is important for penetration. It is generally of the order of 20°.

LIGHTER TINED IMPLEMENTS

Tined implements for secondary cultivations follow the same general principles. The most important group are the spring-tined harrows, available in widths to suit virtually any tractor size, linkage mounted and often with hydraulically or manually folded 'wing' extensions for easy transport. They are used particularly for cereals where some cloddiness of seedbed is acceptable, but rarely for sugar beet and potatoes where the tendency to pull up clods makes them undesirable (fig. 15.3). In this case vertical straight tined harrows, commonly known as Dutch harrows, are used (photos 10.1 and 10.2).

Tines with a reverse slope are occasionally used in conjunction with other implements, notably the spring tine. Once dry soil has been disturbed the clods can be difficult to break. Implements that lift or are neutral in action are often of little value because they stir or disturb the clods without breaking them. The alternatives, then, include backward facing tines, which press downwards on the clods to crush or cut them. Disc harrows and rolls function in a similar manner.

Plate 10.1   Straight-tined harrow, fully mounted and 9 m wide. Sections are controlled by a closed hydraulic system
(John Wilder Engineering Ltd.)

Plate 10.2   Straight-tined harrow, with front board and rear
crumbler for clod breaking   (Cousins of Emneth)

CULTIVATOR FRAMES AND TOOLBARS

Frame design and tine placing are important. The frame should be
strong enough to withstand all reasonable loading without distort-
ing, and with heavy cultivators shear bolts in the tine fitting are an
advantage. The most important point is that the frame should be
designed so that any trash or soil build-up at the top of the tines
should be able to clear through the frame. The point was made in
Chapter 7 that loading or pressure on the soil tends to increase its
strength. When severe jamming up of rubbish occurs under a
cultivator frame, draught is likely to be higher for the same reason.
High soil loading by incorrectly placed depth wheels can similarly
cause draught increases.

DRAUGHT AND EFFECT ON SOIL

As far as is known there is every advantage in keeping draught of
cultivators down by correct frame design and tine approach angle.
There is nothing to suggest that the higher draught of the vertical

tine in particular does anything better in the way of cultivation than lower draught designs: the reverse is likely in many cases.

VIBRATING TINES*

Cultivators that vibrate through contact with the soil (known as freely vibrating cultivators), as opposed to mechanically vibrated tools, depend on high forward speeds for optimum action. Spring tine harrows need to move at 8 km/hr or faster and are then likely to give decreased cloddiness, better cleaning off of weeds, less soil sticking to them, and lower draught than rigid designs. The draught advantage may be as much as 22–55 per cent in high clay content soils but very little or nothing in sands.

Power-vibrated tines for deeper cultivation work have been used on a limited scale for some years. The vibration is normally backwards and forwards, and while total power application is probably not much reduced, the advantage of power take-off transmission for heavy subsoiling work is considerable (photo 10.3). Draught reduction has been found to occur when the oscillation rate is such that the soil is broken into sections shorter than that which would normally be pushed ahead of the same tine when not oscillated. Draught reduction as great as 80 per cent has been recorded, but at this level the total power input (draught plus pto power) was greater than for a rigid tine. Maximum efficiency probably occurs at 40–50 per cent draught reduction. The important point is that power vibrated tines allow deep subsoiling without the use of a very heavy wheeled tractor or crawler. Improved uniformity of soil fracturing because shear planes tend to develop with each oscillation and because the operator can control the rate of oscillation is also claimed.

RIGID SUBSOIL TINES

While it has been suggested that lower draughts can be achieved by angled cultivator tines or legs, this idea is not practicable for deep subsoiling in dry clay soils because of the mechanical strength problems involved in the design of such an implement. The usual design is a vertical leg with a replaceable shin on its

* Vibrating tines are distinct from reciprocating tines which are discussed in the following chapter.

Plate 10.3   Power-vibrated cultivator   (F.W. McConnel Ltd.)

leading edge to cut through the soil. Share design is important to avoid pushing a cone of soil ahead of the metal and so causing increased draught.

GROUND-DRIVEN HARROWS

The rotary or orbital harrow (photo 10.4) has tines which are set on the circumference of frames which are free to rotate, the rotation being provided by some of the tines always being in the soil and in line with the direction of travel of the tractor. They are of limited commercial importance.

DISCS

Discs may be used for either primary or secondary cultivation, depending on the size and weight of the discs. Energy demand is low at about 95 MJ/ha for once-over shallow primary cultivation

Plate 10.4   Illustrating the performance of one type of ground-driven cultivator – the Sampo   (Falcon, Stafford)

for cereals and 27 MJ/ha for secondary operations. Energy demand does not rise steeply with increased forward speed.

Primary cultivation work rates can be high at up to 80 ha/day with a large tractor, and 'down' time is extremely low with as much as 1500 ha being cultivated by one implement without major replacement of parts. The entire soil surface is disturbed in one pass. It has been shown experimentally that maximum crop yields can be obtained with this system of cultivation, but there is reason to accept that discs are potentially soil compactors and extreme care needs to be taken in their use in heavy, moist and previously compacted soils.

Discs in disc ploughs are mounted singly and are angled to the vertical and to the direction of travel. They have limited use in Britain, but may gain favour with the increasing need to bury and incorporate chopped straw. (See page 141.)

Discs in the form of disc harrows, as used for primary and secondary cultivation, are grouped together on axles and can be adjusted only for angle to the direction of travel. The range normally available is 0–30° in tandem and offset implements, and 35–60° in vertical disc ploughs.

Fig. 10.6 illustrates the main arrangements for vertical discs. Tandem discs allow adjustment of front and rear gangs separately within the frame and are the popular arrangement. Offset discs consist of two straight rows angled at up to 30°. They get their name from the ease with which they may be operated offset from the centre line of the tractor. Single acting discs, consisting of a single row of discs angled at 35°–60° to the direction of travel, are not commercially important.

Disc shape is generally concave, and the cutting edge may be plain or scalloped. The former gives better pulverisation and covering; the latter better penetration. The entire implement may be fitted with discs of one type, or with the front row discs scalloped and the second row plain. The reverse, with the front row plain, is less commonly used. Alternative disc shapes are conical, which is said to require less force for penetration, or convex centre, which is said to give more soil inversion.

Discs for primary cultivation are generally 610 mm (24 in) or 660 mm (26 in) diameter and carry weight in the range of 35–150 kg/disc. There are models available at well over 760 mm (30 in) diameter carrying weight up to 400 kg/disc. These should be considered as tools for reclamation rather than routine primary tillage.

For secondary cultivation, diameters of 500 mm (20 in) or less may be used with weights as low as 25 kg/disc.

Disc spacing is usually about 230 mm (9 in), and in tandem discs may be 270 mm (11 in) in front and 230 mm (9 in) behind to combine the benefits of good clearance in surface trash and maximum soil disturbance.

ROLLS

Ribbed and plain rollers act downwards in the manner of disc harrows and backward sloping tines. Cambridge or ring rolls are offered in the range 170–350 kg per metre of width, while plain rolls may be offered in weights considerably above this level. The heaviest of plain rollers are intended for pushing down large stones rather than a cultivation treatment as such.

The most important aspect of roller use is the forward speed employed. Effect falls off rapidly as speed increases and 3 km/hr is probably fast enough for successful use. The falling off in

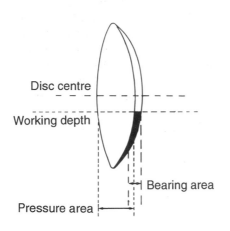

Disc centre

Working depth

Bearing area

Pressure area

(A) Disc bearing and pressure area
(after McCreery 1958)

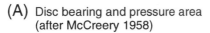

(B) Single-acting discs or vertical disc
ploughs, which operate at 35–60°
to the line of travel

Direction of travel

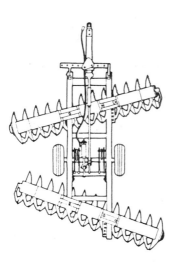

(C) Offset discs, set at up to 30° to
the line of travel

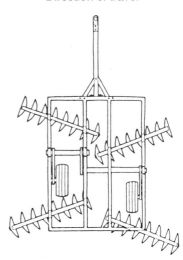

(D) Tandem or double offset discs,
angled from the centre of the
implement

*Figure 10.6   Diagrammatic representation of common forms of disc
harrow* (Instut Techniques des Céréales et des Fourrages, Paris)

percentage of effect with speed increase is shown in fig. 10.7.

It was at one time thought that the reason for reduced perfor-mance was that the roller bounced over the soil surface, but it has been demonstrated that the actual time the roller is on the ground is vital and a roller designed to eliminate bounce would not be an advantage. The compacting effect of most rollers is not great, and at the weights quoted will generally not give much consolidation below 50 mm depth. Cambridge rolls are now available in widths up to 12 m, with hydraulic folding or end-towing facilities to aid between-field movement.

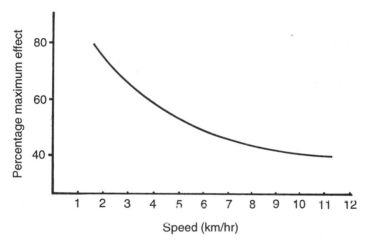

Figure 10.7   Illustrating the rapid decline in the effect of the roller as speed increases   (ASAE)

MULTIPLE IMPLEMENTS

A full load for most tractors can be provided by chisel ploughs, heavy cultivators and spring tine harrows. Disc harrows are more difficult because under most field conditions draught is light, and an implement wide enough to fully utilise the power of the larger tractors is large and expensive. In the lighter types of tined harrow a full load is generally only possible with the help of multiple hitches. While these are available, often in automatic coupling form, they are not widely used in Britain (photo 10.5).

The use of combined tilth-producing implements is more often aimed at producing the desired finish in one pass than making up

Plate 10.5   Multiple hitching of heavy discs   (Parmiter)

a full load for the tractor. There are many examples of this type of equipment available and they are largely aimed at the finer type of seedbed. They may be built into one unit, with clod crumblers and a levelling bar in one frame with a tined harrow (photo 10.6), or the combination may be made up of separate implements. For

Plate 10.6   Spring-tined harrow combined with backward-sloped crumbler tines   (Agrolux-Ransomes)

example, a spring tined harrow which tends to raise clods is followed by a rotary crumbler harrow with tines sloping backwards. Then, as clods are lifted, they are broken and firmed down by the following equipment.

Further useful combinations of powered and non-powered implements are discussed in Chapter 11 with information on power requirements and implement tractor matching and work rates set out in Chapter 12.

EQUIPMENT FOR DEEPER SOIL WORKING

Mole Ploughs

The mole plough is required to produce a permanent channel in the soil at a constant gradient. The smooth channel is produced by a sharpened cartridge followed by a large expander which compacts and deforms the plastic clay to form the channel (fig. 10.8). Depth control is desirably by the soil forces on the cartridge and should not depend on large wheels or on the beam skidding over the soil surface. The main objections to the skidding action of the beam are high power requirement, the likelihood of trash blockages and the inability to produce an inclined mole in a horizontal soil surface. These have been overcome by the 'floating beam' design, which gives up to 40 per cent draught reduction,

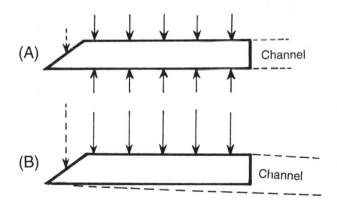

Figure 10.8   A long beam is required on the mole plough to allow soil forces to control grade effectively. At (a) the mole runs correctly and the forces above and below, as shown by arrows, are equal. At (b) it is tending to run out of grade and will be corrected by the large downwards forces, and absence of upward force (E.C. Childs 1942)

the ability to work through some trash, a stabilising effect which reduces localised grade variation, and the means of controlling working depth by hydraulic control of the hitch point which is itself set back from the tractor drawbar. Adjustment of the rake angle of the leg is by steel wedge, or occasionally by screw adjustment. The long beam is essential to allow soil forces to act on the cartridge for control of gradient.

Mole ploughs are mounted on steel wheels or rubber tyres and the lift mechanism may be mechanical or hydraulic. The beam may have a raised end section to avoid surcharging the soil where it lifts and comes into contact with a straight beam under some conditions.

Subsoilers

Traditional subsoilers are designed to work in the range 375–500 mm deep and generally consist of a substantial leg with a replaceable shin on its leading edge, with a 75 mm wide metal foot protected by a chisel share (photo 10.7). The draught of this equipment is substantial (see Chapter 12) and some care is required in its use (see Chapter 13). Crawler tractors are widely

Plate 10.7   Subsoiler with wings and leading tines

used with subsoilers, two subsoil tines being considered acceptable loading in most cases.

The addition of wings to the sides of the subsoiler foot can increase the amount of subsoil loosening by 3–4 times compared with the conventional design, for a draught increase of only 20–30 per cent (photo 10.8). The exact position and dimensions of the wings will depend largely on the make of subsoiler being modified, but the general principles are:

- The wings should be within the central part of the foot.
- Care should be taken not to obstruct the pin holding the replaceable shin in position.
- The overall span of the wings should be about 300 mm.
- The 'lift' of the wings should be 75–100 mm.
- The cutting edge of the wing should be horizontal, with the backward sweep angle at 30–50° and the lift angle at 20–30°.
- The bottom cutting edge should be about 25 mm above the working depth of the chisel share.

Plate 10.8   Subsoiler detail showing leading disc, replaceable shin and wings. Note that the assembly has some freedom of movement on the main frame   (N.J. Cooper)

Plate 10.9   Detail of the 'Flatlift' subsoiler

Plate 10.10   The 'Flatlift' in use; note the large depth wheels

More effective disturbance of the soil can be achieved by placing leading tines ahead of the subsoiler. These shallow tines do not increase total draught and frequently reduce it. The principle is often taken further to form a three-stage system on one tool frame, where A blades at 50–75 mm depth are followed by tines at about 180 mm, which are in turn followed by subsoil tines. This principle, known as the 'Progressive principle', achieves substantial soil disturbances in one high draught pass through the soil.

SHALLOWER SOIL LOOSENING

Under circumstances where loosening of the upper soil layers is required, possibly after minimal cultivation, direct drilling or intensive grazing, special-purpose equipment can be used. The requirement is for complete disturbance of the soil to the depth required (seldom in excess of 300 mm) with little or no loss of surface condition. This is obtained through the use of large 'wings' and close spacing of tines. The 'Flatlift', as shown in photos 10.9 and 10.10, is a successful example, being suitable for use with tractors of the order of 80 kW and above for full tractor-width operation.

CHAPTER 11

CULTIVATION IMPLEMENTS: Powered

The use of power from the tractor power take-off rather than by way of the tyres and linkage or drawbar is a solution to some of the problems involved in matching implements to high-powered tractors. There is little or no draught required, so that problems of wheel slip and heavy ballasting do not occur. More work can be done on the soil at one pass and with one set of wheelings, and the full power of the tractor can be used with an implement of manageable width. Powered equipment offers the possibility of quicker seedbed preparation and therefore better timing of cultivation but, at the same time, it is more open to misuse in wet conditions with consequent soil damage.

ROTARY CULTIVATORS

Rotary cultivators generally work on a horizontal shaft, and less commonly on a vertical or series of vertical shafts. They require more power for a given volume of soil disturbed* in primary cultivation than the draught implements, but not necessarily more power for a given degree of soil break-up. The power used for acceleration of soil particles, much of which is lost for cultivation purposes, is much higher than in mouldboard ploughing. Against this has to be set the inefficiency of power transmission through rubber tyres, especially under loose surface conditions. Table 11.1 sets out a comparison of power use for a mouldboard plough, spading machine and rotary cultivator in silty clay at normal cultivation moisture content. The peripheral speed of the rotary

* This is occasionally termed a 'higher specific power requirement'.

Table 11.1   Performance of spading machine, rotary cultivator and mouldboard plough in silty clay loam

Forward speed km/hr	Total energy kJ/m³	Equivalent energy kJ/m³ (a)	Mean clod size mm	Energy MJ/ha 108 mm deep
Spading machine				
0.9	60.33	47.11	280	107
1.8	65.02	34.14	550	115
2.1	60.57	22.36	830	107
Rotary cultivator				
0.8	252.57	48.31	240	448
2.4	150.34	53.00	220	297
4.2	126.40	43.95	310	224
Mouldboard plough				
3.6	45.97	36.87	470	82
6.9	69.91	35.34	540	124
10.4	94.09	35.34	550	167

(a) Equivalent energy is the amount of energy required to produce a comparable clod by a standard laboratory method. Therefore a high figure indicates small clods.
Source: A. W. Cooper 1963

cultivator was low at 3.8 m/s and would therefore tend to show it at its best from the efficiency point of view.

POWER REQUIREMENTS COMPARED

It will be seen in Table 11.1 that the spading machine and the mouldboard plough require much the same amount of energy to give the same mean clod size, and therefore similar energy to cultivate a hectare. The higher power requirement of the rotary machine is shown. It has been suggested that the rotary cultivator compared with a mouldboard plough plus traditional trailed equipment to produce the same seedbed generally takes up power in the ratio 1.5 to 1.

The degree of soil break-up achieved with the rotary cultivator depends on the ratio of rotor speed to forward speed, the magnitude of the speeds, and the shape and arrangement of the tines or blades. Rotation is normally in the same direction as the tractor tyres but can be in the opposite direction, and the form of slice differs according to whether it is cut in the usual surface-downwards direction or the reverse. The relationship between length of cut, forward speed and rotor speed for a three-blade rotor is shown in fig. 11.1.

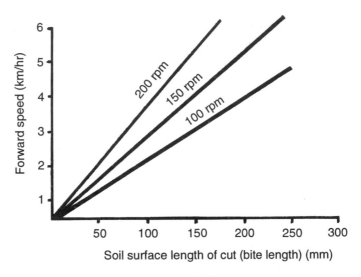

Figure 11.1   Length of cut of a three-blade rotor   (ASAE)

Control of Degree of Pulverisation

The extent of soil pulverisation is reduced by reducing rotor speed, increasing forward speed, using the two- rather than three-blade layout (fig. 11.2), raising the shield or by shallower

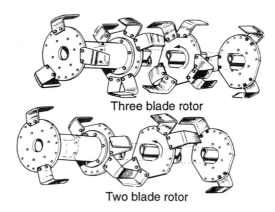

Three blade rotor

Two blade rotor

Figure 11.2   Rotor blade layout   (Howard Machinery Ltd.)

working. Deeper working increases pulverisation even though the size of slice may be greater. This is because it tends to increase recirculation of soil within the machine. High rotor speeds tend to give greater pulverisation, even for a given size of cut (that is, with the forward speed increased to give the same ratio) because of greater recirculation and greater impact on the hood (fig. 11.3).

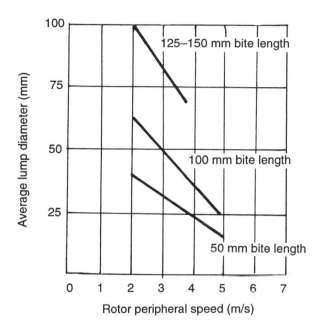

Figure 11.3 Showing how an increase in rotor speed gives more pulverisation regardless of bite length

The high power requirement of the rotary cultivator is particularly concerned with high peripheral speeds and therefore high acceleration losses, as well as larger cutting surfaces as compared with the mouldboard plough. It follows that power requirement can be reduced by rearrangement of blades and optimum forward and peripheral speeds.

Figs. 11.4 and 11.5 illustrate the effect that decreasing soil surface length of cut ('bite' length) has on power requirement and the effect of depth of work on specific power requirement.

Blade Shape

The three main types are the hoe, the pick tine and the knife or

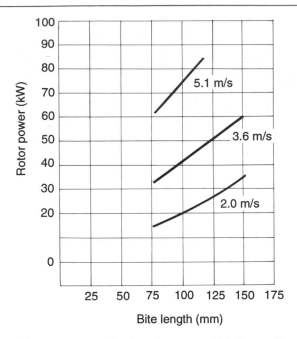

Figure 11.4   *The power required at the rotor at high, medium and low tine-tip speeds (measured in m/s)*

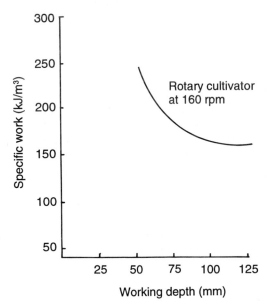

Figure 11.5   *Reduction in specific work with increasing depth*

slicer (fig. 11.6). The hoe has been developed for near universal use and performs well from the point of view of degree of pulverisation, power requirement, mixing of surface residues or manure, chopping of surface material and freedom from blockage. The other forms of tine may be better than the hoe for specific purposes such as cutting up matted pastures, or breaking hard and compacted soil. Pick tines tend to produce a rougher tilth than hoe tines. One characteristic which makes the hoe less desirable than the pick tine is its greater tendency to smear the soil at the depth of cultivation in some conditions.

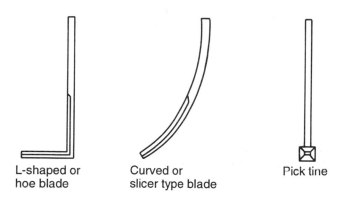

L-shaped or   Curved or   Pick tine
hoe blade   slicer type blade

Figure 11.6   The three basic types of rotary cultivator tine

Blade Layout
The layout of blades on the rotor is important for achieving the tilth required and keeping power requirement down to the minimum practicable. Fig. 11.2 shows the two most important arrangements in farm use. The flanges on the horizontal shaft are normally drilled to accept either six blades in three pairs (the 'three-blade' arrangement) or four blades in two pairs. The latter arrangement gives a rougher tilth, lower power requirement and higher working rates.

Pick tines are fitted on flanges spaced at 125 mm on the shaft instead of the usual 250 mm, the reason being that they are narrow and would not fully disturb the soil at the full spacing width.

Rowcrop Versions
There are various rowcrop blade formations possible to allow work

between such crops as strawberries, grass for seed and maize. Crop guards are frequently used with this equipment to shield the crop from disturbed soil. Rowcrop machines can be reset to the standard flange spacing for full width cultivation without difficulty.

Effect of the Shield

The rear shield is an important part of the rotary cultivator, the height at which it is set having a dramatic effect on tilth. It should be set 'down' for severe pulverisation and 'up' for a cloddy tilth or where the soil is moist and liable to damage if over-cultivated.

The effect of the shield when right down is to recirculate the clods through the rotor, as well as to act as a levelling board on the soil and a barrier against which clods are broken (photo 11.1).

Choice of Gears

The gearbox of the rotary cultivator is important, and improvements in the range of speeds available and ease of selection have been major advances in machine design. Full use of the gearbox is essential if the important property of this system of cultivation— that the operator can choose the type of tilth he requires—is to

Plate 11.1   Rotary cultivator with shield in raised position
(Howard Machinery Ltd.)

be exploited. Blades and rotors are designed for a particular bite length or distance between each cut. If this is considerably exceeded, an inefficient and undesirable action known as trowelling occurs. At less than the optimum cut length, efficiency also falls. If gearing is good, forward speed can be matched to implement size and available tractor power.

Special Applications

Recent developments of this type of rotary cultivator include spiked rotors for the production of fine seedbed tilths, blades arranged to cut slots for direct drilling from a mounted seedbox into unploughed land, and seedboxes mounted over the rotary cultivator for drilling into land that has received a minimum amount of cultivation. Rotary cultivators have been used for light cultivations aimed at fragmenting 'couch' (*Agropyron repens* and *Agrostis* species). For this purpose the two-blade rotor arrangement is useful because it allows high-speed working.

Power requirements and operating speeds are suggested in Chapter 12.

SPADING MACHINES

Spading machines or rotary diggers operate at lower peripheral speeds than rotary cultivators, take a larger bite, and therefore produce a rough tilth (photo 11.2). Specific energy requirement, as indicated in Table 11.1, is nearer that of the mouldboard plough than the rotary cultivator. The shape and speed of the blades of the spading machine are such that they are not self-cleaning and the slice has to be tipped off mechanically. The spade-shaped blades are arranged in sets of three on the rotor. As the rotor turns, the spades are held square to the direction of travel for about two-thirds of the revolution. During the remaining one-third they turn through 90° and back again to tip off the section of dug soil.

This system is efficient in power use, but calls for a complicated and expensive machine. Work rate is not high because forward speed has to be low in conjunction with the low spade speed, and a very wide implement would be needed to utilise the power from a medium tractor under most conditions. This implement pushes hard behind the tractor, so no draught is required, and has a particularly gentle action on the soil. It is particularly valuable in intensive arable farming in silt areas.

There are generally coulters fitted to make a cut between each

Plate 11.2   A form of spading machine, which is available up to 3 m width, can incorporate chopped or unchopped straw, and can alleviate soil compaction   (MFJ-Imants)

set of spades and draught power for these is more than provided for by the pushing action of the rotor. Working depth is controlled by a skid, and available working widths are 1–3 metres.

RECIPROCATING HARROWS

Reciprocating harrows have been developed for rapid tilth production after primary cultivation and are distinct from vibrating tined equipment. They consist of from two to four parallel bars with vertical spikes fitted at intervals of about 175 mm (photo 11.3).

The bars move from side to side under pto power at up to 500 cycles per minute. Working depth is usually up to 180 mm and can, where necessary, be controlled by skids or by adjustment of the spikes within the bars. Working width can be up to 5 m for use with large tractors and forward speeds up to 11 km/hr.

The action of reciprocating harrows can with some designs cause severe vibration of the tractor. Ridging of the soil at each

Plate 11.3   Reciprocating harrow   (Vicon Ltd.)

side of the implement can be a problem and combs or anti-ridging plates are sometimes fitted. Crumblers are frequently fitted behind the harrows, and some combined seeders and reciprocating harrows are in use.

Effect on Tilth

Where more than two bars are used, the action usually intensifies from front to rear with each successive bar, either by an increase in the amount of sideways movement (100, 225, 350 and 500 mm respectively, for example) or by increasing the depth of tines.

In some designs tine numbers can be changed to give tilth variation, although this more often depends only on pto and tractor forward speed. Tines can also be removed for rowcrop work.

Reciprocating harrows are a valuable tool for tilth production, especially for root crops, and can work at high speeds without bringing raw clods to the surface. They are potential compactors by virtue of the way they shake down and rearrange soil particles, however, and over-use can be dangerous.

POWERED ROTARY HARROWS

The rotating harrow is used in conditions similar to those favoured for reciprocating harrows and is in many cases a direct competitor (photo 11.4). It consists of a series of contra-rotating rotors, each with two vertical tines or spikes, mounted on a single bar. The rotors are gear driven at around 250 rpm, and the ratio of rotor to pto speed may be fixed or changeable in several stages. Tilth produced is determined through forward speed and rotor speed combination. Depth control is by the following crumbler which also breaks surface clods. Working depths up to 250 mm are possible, and there is no tendency to bring stones or clods to the surface. The rotating harrow can be used for rowcrop work with some of the rotors removed.

Width of work is 1.5 to 8.0 m, and speeds of up to 6 km/hr are common. The maximum width and speed in heavier soils requires a very large tractor.

As with reciprocating harrows, there is need for side plates to take down the ridge formed between bouts. All forms of powered

Plate 11.4   Powered rotary harrow – the 'Roterra'   (Lely Ltd.)

harrows are at risk of damage through blockage by stones and a slip clutch may be needed.

COMBINED IMPLEMENTS

Combined powered and non-powered implements have appeared in many forms. It has been suggested that the non-powered element could ease up the soil along natural planes of weakness, to be followed by the shattering effect of the powered element. In practice the reverse system is in use, where subsoil tines follow the rotary cultivator. Use is then made of the draught developed by the pushing action of the rotor, but there is probably not any saving in total power requirement. The advantage is a saving of a further trip over the field, with draught and wheel slip involved for the tining operation.

Mouldboard ploughs have been combined with rotary cultivators without much commercial success. The most important combination implements are the power harrows and rotary cultivators which are followed by various forms of passive crumblers and clod crushers.

IMPLEMENTS FOR STRAW INCORPORATION

Choice of implements for the incorporation of chopped straw will vary with soil moisture and texture, weed growth, degree of lodging of the previous crop, the date on which the work can take place, and the identity of the following crop. It is unlikely that a set pattern of operations will be optimum for an entire farm in successive years. Chopping of the straw needs to be to average lengths of 50–100 mm, and uniform spreading of both straw and chaff behind the harvester is important. Stubble length is not critical, but is best cut short for discs and tines, and longer for mouldboard ploughing.

The mouldboard plough is the leading implement for straw incorporation and where it is already in satisfactory use where straw has been cleared, it is likely to perform well in chopped straw. The plough is the most effective and economical tool to use on light to medium soils. In the medium to heavy range a combination of plough with tines and discs is likely to be the best. There are no special requirements for ploughs working in chopped straw. Under-beam clearance should be about 700 mm, but need not be the very high clearance offered by some manufacturers. Furrow widths of 350 mm work well, but this is not critical. Skims can function well in dry straw, and the commonly used plough body shapes are satisfactory. The suggestion above that longer stubble is best handled with the plough rather than other implements is because the stubble tends to hold the chopped straw for the plough in comb-like fashion and limits blockages. Extensions to the mouldboards, known as trashboards, have been used for ploughing-in straw but have not met with particular success.

Swing-beam or square ploughs (page 141) have given good burial and mixing of chopped straw in free flowing soils, with some advantage in cost and power requirement, but may not fully replace the traditional mouldboard under wet and heavy land conditions.

Tines and discs provide satisfactory alternatives to the plough on heavier land. In dry seasons moisture loss may be less under non-ploughing systems. The guidelines are that the stubble should be short, consolidation should be adequate, and care should be taken to avoid grass-weed build-up, if necessary by rotational ploughing. The favoured equipment is based on rigid tines for good penetration, in some cases in combination with rotary (powered) cultivators.

CHAPTER 12

POWER REQUIREMENTS AND WORK RATES

Stationary tractor tests give an indication of engine power, fuel consumption and maximum torque, which are of value to the farmer. Power take-off figures are quoted, with a range of governor settings which include maximum power. Tractors from the production line will obviously vary in performance to some extent and the range of power output found for one model may be 10 per cent.

The efficiency of modern diesels is such that fuel consumption (generally quoted as g/kWhr or kWh/l) does not vary widely but is of interest since it can be used in the field to get an indication of how fully a tractor is being utilised.

Torque figures give an indication of pulling characteristics but are not relevant to this discussion.

DRAWBAR POWER FIGURES

Drawbar performance tests are aimed at comparing the drawbar performances of various tractors under standard conditions. Since field conditions cannot be standardised, the tests are carried out on concrete or tarmacadam tracks. This gives the best performance that could be expected from these tractors and will be better than field performance. Maximum pull figures vary according to track conditions by as much as 10 per cent, but the more useful figure of maximum drawbar horsepower is more constant at about 5 per cent variation. Crawlers are tested on clay soil of quoted shear strength.

175

TRACTOR DRAUGHT

Maximum draught on concrete or tarmacadam is of the order of 0.8–1.0 times the weight on the driving wheels. That is, a co-efficient of traction, as defined in Chapter 7, of 0.8–1.0. Under field conditions this is seldom more than 0.5, and over an autumn's work on stubble it might average 0.4. Crawlers, because of the large section of soil sheared by the grousers, normally develop a maximum pull considerably greater than their own weight.

DRAUGHT AND POWER IN THE FIELD

Maximum draught on artificial tracks is normally limited by wheel slip rather than engine power. This is also the case in the field where traction is normally poorer. Therefore if full use is to be made of engine power, speeds in the 5–8 km/hr range have to be used so that the power can be taken up at reasonable draught, giving low slip levels.

In practice the power available for draught work at the rear end of a wheeled tractor is not likely to be more than 60 per cent of quoted engine power, and for a crawler just over 70 per cent of quoted engine power.

The kilowatt is defined as 1 kN metre per second. Therefore power can be calculated from the formula:

Approximate kW = Implement speed (km/hr) × Draught (kN) × 0.28

The approximate answer can more easily be taken from fig. 12.1. A straight edge placed across the page will give the relationship between power, speed and draught. For example, 17 kW at 6 km/hr is equivalent to a draught of slightly more than 10 kN. Therefore, if a light cultivator has a draught of about 10 kN and it should be drawn at 6 km/hr, 17 kW will be needed. This can be converted to engine power as follows:

$$\text{Engine power} = 17 \times \frac{100}{60} = 28 \text{ kW}$$

(Based on 60% of engine power available for draught)

Power requirements for cultivation vary widely with soil conditions and depth of cultivation. Table 12.1 should only be used as a general guide and could well be modified in the light of experience on individual farms.

Table 12.1   Power and draught requirement for cultivation

Machine	Working width	Pto power	Draught
Ploughing (per furrow)	250–400 mm		12.4 kN/m* (range 3–11 N/cm² furrow section)
Chisel ploughing	2–5 m		7.3 kN/m
Primary discing	2–10 m		7.5 kN/m
Disc harrowing	2–10 m		3.6 kN/m
Medium tined cultivation	2–7 m		3.5 kN/m
Spring tine harrows	2–8 m		2.6 kN/m
Seed harrows	3–12 m		1.5 kN/m
Cambridge rolls	2.5–12 m		1.5 kN/m
Drilling (corn)	2.5–6 m		2.9 kN/m
Rotary cultivation	1.2–4.5 m	24 kW/m width	0–1.8 kN/m
Reciprocating harrow	2.5–5 m	7 kW/m width	2.5 kN/m
Rotary harrow	2.5–6 m	12 kW/m width	2.5 kN/m
Mole ploughing	Single mole		Up to 44 kN
Subsoiling clay soil 450 mm deep	1.2 m per tine		Up to 17 kN per tine

* 1 kN/m width of implement is equivalent to an energy input of 10 MJ/ha.

The purpose of the table is to give an idea of the matching width of implement for a known tractor with the help of the ready reckoner in fig. 12.1.

RATE OF WORK

Three terms are commonly used:

Spot Rate
Spot rate of work is the actual rate of work as the implement crosses the field. It does not include turning time, filling time for drills or any other lost time. Spot rate is given by the formula:

$$\text{ha per hour} = \frac{\text{width of work (m)} \times \text{speed (km/hr)}}{10}$$

Net Rate
Net rate of work includes turns but no other stoppages. It might typically be a record of three or four uninterrupted bouts across the

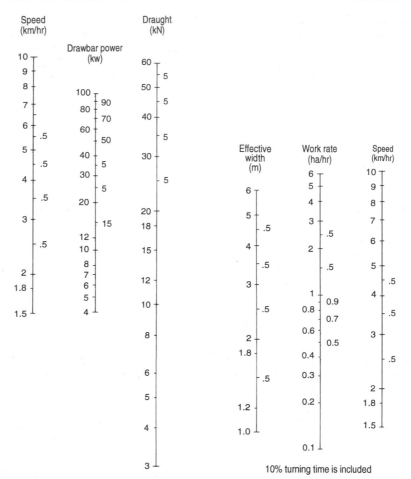

Figure 12.1   Nomograph to aid calculation of drawbar power

Figure 12.2   Nomograph to aid calculation of rate of work

field, and can be calculated from the spot rate by substituting either 11 or 12 in place of 10 in the formula. Fig. 12.2, a ready reckoner for implement width, speed and workrate, is based on the '11' factor in the formula, which allows about 10 per cent for turning.

Overall Rate

Overall work rate is a further reduction to allow for filling, blockages, adjustments, and other stoppages. This is generally calculated by applying an efficiency factor, expressed as a percentage, to the net rate (Table 12.2).

Table 12.2   Forward speed and field efficiency

	Typical forward speed km/hr	Field efficiency %
Ploughing	5.6	80
Chisel ploughing	6.4	80
Disc harrowing	8.0	85
Medium tined cultivation	8.0	85
Spring tine harrows	9.7	85
Seed harrows	8.0	85
Cambridge rolls	4.0	85
Drilling (corn)	9.7	65
Rotary cultivation	4.0	85
Reciprocating and rotary harrows	6.4	85
Mole ploughing	3.2	74
Deep subsoiling	4.0	80

These figures should again only be used as a general guide.

FIELD EFFICIENCY

Field efficiency is defined as the percentage of total time the implement is doing useful work. It is particularly important for operations such as drilling where the organisation of seed and fertiliser hopper filling can have a major effect on the percentage achieved. One point to note is that the faster the drill moves across the field, the lower will be the efficiency factor for a given filling rate. This is because the drill will be spending less time actually in the soil for each re-fill, and is of course not reflected in poorer output per day.

Example
A tractor rated at 100 kW is likely to give 60 per cent usable drawbar horsepower, say 60 kW.

We wish to chisel plough at 7 km/hr, so from fig. 12.1 the draught available is 31 kN.

From Table 12.1 this gives an implement width of $\frac{31}{7.3}$ or 4.2 m.

The net rate of work, from fig. 12.2 for 4.2 m at 7 km/hr is 2.7 ha/hr.

The efficiency factor suggested in Table 12.2 is 80 per cent, to give an overall workrate of 80% × 2.7 = 2.2 ha/hr.

MAXIMUM DRAUGHT AVAILABLE

Maximum draughts are of interest particularly for moling and subsoiling. It has been stated that crawlers can develop draughts equivalent to or greater than their own weight. A rough indication of performance is thus available.

Wheeled tractors are likely to give the following maxima under dry autumn field conditions:

    Two-wheel drive 45 kW   3250 kg
    Two-wheel drive 56 kW   3750 kg
    Four-wheel drive 75 kW   4500 kg

Under good traction conditions wheeled tractors will develop draughts high enough to damage their own transmission systems and some caution is desirable.

EFFECTIVENESS OF USE OF FARM TRACTORS

In these proposed systems of reckoning it has been assumed that tractors are used to somewhere near their full capacity. In practice this seldom occurs and survey work on fuel use in tractors suggests that they are operated at between 50 and 70 per cent of what they could achieve (fig. 12.3). Fuel consumption is a useful check on the farm for tractor output. One litre of diesel oil an hour gives roughly 3.3 kW at the engine, so that a fully used 75 kW engine uses about 23 litres per hour. The best that could reasonably be achieved in the field is 18 litres per hour usage, and the survey average for this size of tractor is well below 14 litres.

There are several known reasons for under-usage of tractors. The main ones are poor implement matching, driver discomfort through noise and vibration, and inadequate gear ranges. All these factors are the subject of continuous improvement from tractor manufacturers. In all field work the output from any one machine will vary widely with different operators.

FIELD SIZE

Small fields involve more turns and loss of time through short runs. Efficiency improves with larger field sizes up to about 16–20 ha but seldom beyond that point.

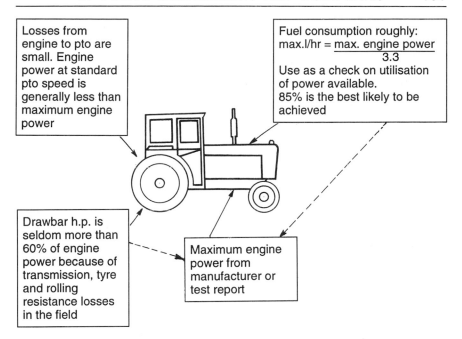

Figure 12.3   Tractor power

CHAPTER 13

SUBSOIL CULTIVATIONS

Several references have already been made in earlier chapters to the need for well fissured subsoils, which will allow free movement of drainage water and free growth of roots. The means of achieving this condition, where it is absent, are considered in this chapter.

TYPES OF SUBSOIL CULTIVATIONS

There are six types of subsoil cultivation used in British arable farming; unfortunately names for these vary in different parts of the country and to avoid confusion the terms used in this chapter are defined below.

Plough Subsoiling
Some loosening below the plough furrow is achieved with a tine or blade fitted to the plough. Usually this tine extends no more than 75–150 mm below the furrow bottom in which the tractor wheel will next run. Sometimes the blade has a shoe or bullet at the base to improve the busting action. These devices increase draught significantly but on lighter land are effective in reducing plough pans. Heavy land is usually too moist at ploughing for effective shattering.

Deep Cultivating
Heavy rigid tine cultivators can generally be used to shatter compacted layers to a depth of about 300 mm. At this depth, providing the power available is sufficient, it is possible to pull tines close enough to completely break the panned layer. A

common fault is to lose depth of operation by attempting to pull more tines than available power will allow.

Subsoiling
Compaction extending deeper than about 300 mm requires the subsoiler. This is generally a single or double bladed implement, with sharpened leading edges, and angled lifting shoes, at the base of the blades (see Chapter 10). Many heavy textured soils have naturally tight compacted subsoils, in which drainage and rooting can be improved by regular deep subsoiling. Subsoiling in clay soils without adequate pipe drainage backfilled with permeable material can do more harm than good (see pages 63 and 193).

Mole Ploughing
The beam mole plough is a trailed implement, which drags a round bullet with expander through the soil, to create a graded drainage channel (see page 67). It is ideal for improving permeability of uniform textured clay soils, thereby allowing wider spacing between pipe drains. At the optimum moisture content for moling, which is moister than for subsoiling, a substantial amount of soil loosening takes place above the mole channel. In general, moling is preferable to subsoiling as a permeability aid on heavy land, except where variability in subsoil texture leads to unstable channels, or where there are severe plough pans.

Deep Ploughing
With deep digging bodies land can be ploughed to 500 mm or more. Although subsoil compaction is eliminated by ploughing to sufficient depth, topsoil organic matter is diluted on all mineral soils, and on heavy soils intractable raw clay is brought to the surface. So the disadvantages usually exceed the advantages, and the practice has rightly been superseded by subsoiling.

Subsoil Raising and Mixing
In certain soils raising of subsoil material and mixing with the topsoil is beneficial. This technique has a very restricted application and so far has been used in the East Anglian Fens on shallow peats over clay or sand subsoils, on shallow clay over peat, and on silty clay soils over sand. The implement used has a broad mixing blade drawn through the soil at an angle of 45° to the horizontal. Subsoil rides up the blade and spills off into the topsoil.

Of the six techniques mentioned only deep cultivating and subsoiling are considered in the remainder of this chapter. One of the earliest subsoilers was built by a Gideon Davies in America in 1845. An early reference to subsoiling in Britain was made by a Mr Johnston, of Penicuik, Midlothian, who was a great believer in 'loosening the subsoil', and who quoted a number of results with oats and swedes to support his beliefs. In arable areas of England, the use of deep subsoiling has now become a routine cultivation on many farms, particularly those with root crops. Although there is no doubt that subsoiling can be beneficial where subsoil compaction is restricting water movement or root penetration, the routine use of a subsoiler on land every third or fourth year, without first establishing the presence of compaction, is difficult to justify except on the basis of reducing risk.

EFFECT OF SUBSOILING ON SOIL PROPERTIES

Soils that have been subsoiled in trials show that there are three beneficial changes in soil properties caused by the deep loosening, all of which are related to increased fissuring. These changes are:

Increased Infiltration
The influence of subsoiling on infiltration rate of soils of several textures has been shown by ADAS soil scientists in the West Midlands (see Table 13.1).

Table 13.1   Increases in infiltration rate after subsoiling

Soil type	Infiltration rate (cm/hr)	
	Not subsoiled	Subsoiled
Clay	0	27
Silt loam	16	27
Loam	111	176
Sandy loam	360	1028

Improvement in soil permeability, whether achieved by moling or subsoiling, is particularly important on clays, and can be just as crucial in intensive grassland farming as it is for arable cropping.

Increased Available Water
In addition to increasing the quantity of larger pores through

which drainage water percolates, subsoiling sometimes adds to the finer porosity which stores water available to roots. This change is very welcome on light textured land in any part of lowland Britain.

Better Root Growth

Improvement in coarse porosity of subsoils after subsoiling not only helps water movement, but allows faster and more extensive root development. In the drier eastern part of England, deep-rooting crops such as sugar beet can take advantage of this improvement, and are able to withstand droughts with less reduction in yield. However, this benefit is not confined to deeper rooting crops as shown in Table 13.2, which gives the distribution of grass roots before and after subsoiling.

Table 13.2 Effect of loosening a panned subsoil on distribution of grass roots

Depth (mm)	Percentage of total grass roots	
	Not subsoiled	Subsoiled
0–100	82.4	69.8
100–200	13.3	13.1
200–300	2.7	8.0
300–400	1.0	5.0
400–500	0.6	2.4
500–600	0.0	1.2
600–650	0.0	0.5

SOIL TYPES AND CONDITIONS THAT BENEFIT FROM SUBSOILING

Subsoiling has been widely practised on many soils with advantages wherever there is a layer in the profile restricting water movement and root growth (photo 13.1). Such layers are formed in a number of ways, either naturally during the very long process of soil formation, or by damage to subsoils from farm implements and tractors in the recent past (photo 6.6). Those soils with naturally compact layers causing slow drainage are more prone to damage from machinery, so that not infrequently both natural and artificial compaction are found in the same soil.

Some soils are much more likely to need sub-soiling than others. In general, the soils most susceptible to subsoil compaction are sands, gravel soils, clays and silts. These soils pack down easily, and subsoiling at regular intervals will usually be necessary. Chalk and limestone soils, and many other well structured medium-textured loams, are much less likely to need subsoiling, particularly if their organic matter level is not too low. Heavy textured land is susceptible to panning, and often responds to a busting operation but deep cracking in dry summers is very beneficial on these soils (photo 6.1).

NATURALLY OCCURRING LAYERS THAT NEED LOOSENING

In very light heath soils, for example, Arne and Winifrith Heaths in Dorset, Bagshot Heath in Surrey, and the Suffolk coastal sand, very strong dark coloured pans of iron and humus are found in the subsoils due to prolonged leaching under heathland vegetation (fig. 13.1A). Provided these pans are not too deep and thick, they can be shattered by a powerful tractor and subsoiler and crop roots are then able to penetrate much deeper into the profile.

Where sand and gravel soils overlie impervious clay, water levels fluctuate in the lighter material, and often strong dark coloured pans consisting mainly of iron and manganese form at the upper water level (fig. 13.1B).

In several of our clay soils that do not have a natural content of lime, clay particles have slowly moved down the soil profile during their formation, causing a heavier textured layer or clay pan in the subsoil. Sometimes this is found in soils that are fairly free draining below the clay pan—the red clay soil on Keuper Marl, for example

Plate 13.1   Soil core pinpoints compacted layer between 200 mm and 300 mm which was affecting growth of barley   (Arable Farmer)

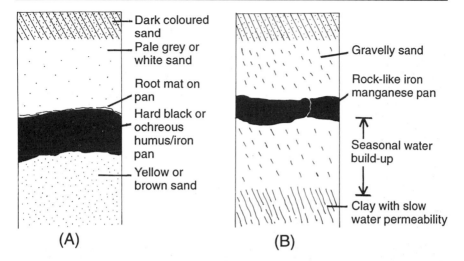

Figure 13.1   Naturally occurring soils with strong 'chemical' pans

(fig. 13.2). In others this clay shift has taken place in a more poorly drained soil, for example, some Lias clays, and many poorly drained sandy boulder clays of east Suffolk and south Norfolk (fig. 13.3). In the former, satisfactory drainage can sometimes be achieved by subsoiling alone, but in the latter pipe drains with permeable back fill over the drains are always necessary, in addition to moling or subsoiling to loosen the clay-enriched layer. In yet other clay soils the permeability becomes less and less with depth, for example, London and Oxford Clays. Mole drainage is preferable to subsoiling on this type of land (photo 13.2).

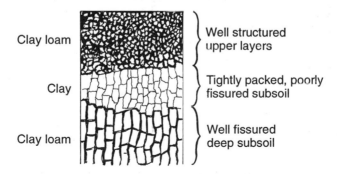

Figure 13.2   Soil in which subsoiling alone can be beneficial

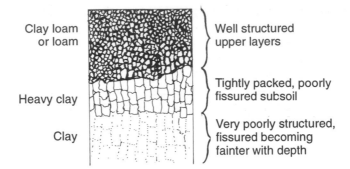

Figure 13.3   Soil in which pipe drains are needed in addition to moling or subsoiling

Plate 13.2   Fissures in a clay soil caused by mole drainer
(Crown copyright)

COMPACTION BY FARM MACHINERY

Compaction caused in subsoils by farm machinery is rarely found deeper than 450 mm, and frequently extends no deeper than 350 mm. Subsoil compaction gradually builds up over a number of years and ploughing can play a major part in this. If land is ploughed wet, and at a similar depth year after year, the furrow bottom becomes smeared over by tractor wheels and by the plough shares. This effect is accentuated by excessive tractor wheel slip, worn plough shares, and by weak structured easily smeared soil. Once a pan is formed the soil is more easily smeared the following year, and a vicious circle is started. Heavily laden trailers and beet harvesters can cause similar damage in the subsoil, when they sink through wet topsoil (photo 13.3). Subsoil pans may hold up water, and in unstable structured soils, slaked material builds up above the compacted layer adding to its thickness.

Plate 13.3   Patches of poor cereal growth associated with soil damage during harvesting of sugar beet   (Crown copyright)

SUBSOILING AND CROP YIELD

Experiments on soils where there is a definite need for subsoiling, of the type described in previous sections, show yield advantages from subsoiling and confirm field experience, but if soil examination is unable to establish the presence of a compacted layer or soil horizon, then crop yield is unlikely to benefit. The actual increase in yield to be expected from subsoiling in a particular situation depends on crop species, weather, and presence of other restrictions to yield. For example, when a sandy loam over cemented sand and gravel was subsoiled in the East Midlands, barley gave a 500 kg/hectare response in 1965, the first year after treatment, but only 250 kg/hectare in the second year after subsoiling. The main reason for this difference was that May and June were much drier in 1965 than in 1966, and the extra water extracted by the deeper rooting crop in the subsoiled area, gave greater benefit in the dry year. Conversely, on heavy land, where subsoiling has been used to improve permeability of the subsoil, larger responses are recorded in wet seasons.

Where cereals and legumes are the main crops grown and the standard of husbandry is good, the need for subsoiling will be less than on farms where root crops are grown, and the risk of autumn damage to the soil is higher (photo 13.3).

SUBSOILING PRACTICE

Having identified a need for subsoiling (see Chapter 15) it is imperative that upward soil heave occurs during subsoiling; sideways deformation is only satisfactory for mole channel formation. Depending on soil type and its condition there is a maximum depth beyond which a satisfactory heave and loosening cannot be achieved. This limiting or critical depth can be increased by widening the leading edge of the subsoil tine, by loosening the surface layers of soil ahead of the subsoiler and by carrying out the operation when the soil is neither too soft and wet nor too hard and dry. It is always necessary when starting in a field to use a spade to check whether the depth of working is giving satisfactory loosening. If the subsoiler is leaving a channel or 'square mole' the toolbar should be raised until maximum loosening is observed in the soil pit. Correct spacing of the subsoiler tines will leave a level soil surface and the use of a disc or knife coulter ahead of the subsoiler leg helps to minimise both surface tearing in grassland and topsoil mixing.

Performance of Subsoilers

The cross-sectional pattern of soil disturbance produced by a subsoiler working above the critical depth is shown in fig. 13.4. A v-shaped zone of soil is broken loose to the surface and horizontal fissures develop on either side of the foot of the implement. Most of the disturbance is in the topsoil and not in the subsoil. The

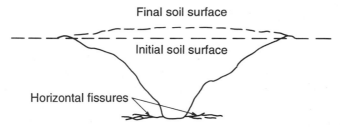

Figure 13.4 Soil disturbance pattern produced by a conventional subsoiler

disturbance can be improved considerably by attaching inclined blades or wings to the side of the foot which are positioned to run in the horizontal cracks (photo 10.10). Although the winged subsoiler has 10–30 per cent extra draught compared with the conventional foot, the subsoil disturbance it creates is two to three times greater (fig. 13.5).

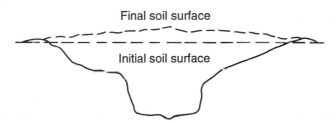

Figure 13.5 Soil disturbance pattern produced by a winged subsoiler

Advantages of Cultivating Before Subsoiling

Reducing the strength of undisturbed surface layers before subsoiling results in increased subsoil disturbance by the subsoiler and an increase in the effective working depth. This can be done by shallow tining or ploughing before subsoiling, or better by attaching shallow tines to the subsoiler frame itself. These tines must be ahead of the subsoiler foot by a distance at least equal to the working depth of the subsoiler and positioned laterally.

Depth of Subsoiling to Loosen Pans

Where the underside of a pan is within the critical working depth, the foot of the subsoiler should not be set more than 40–60 mm below the pan. If the pan is below the critical depth, busting can only be achieved by loosening in two stages of depth.

Soil Moisture and Subsoiling

Subsoiling is most effective when the subsoil is moderately dry. If the soil is very dry, draught is excessive and the soil breaks into very large unfractured blocks; in contrast if the soil is too wet smearing results with much reduced loosening along with the formation of a square channel in the subsoil if the critical depth is exceeded. For sandy soils the moisture range for effective subsoiling is much wider and useful loosening can be achieved in quite wet soil. Compaction in wetter soil can be better relieved when winged subsoilers are preceded by cultivator tines than when the conventional subsoiler is used. No specific guidelines can be given about when it is too wet or too dry for effective subsoiling because too many factors are involved. The best way of finding out is to have trial runs and then to dig a trench across the work; adjust the depth to give maximum loosening and, if this is too little to be acceptable, stop subsoiling.

Frequency of Subsoiling

Subsoiling is required as often as the soil profile shows there is a need; it is not possible to generalise on the frequency of need. Some farmers, by refusing to cultivate or run on land when it is wet, and by encouraging earthworm activity with manure and occasional leys, avoid the need for subsoiling even on easily compacted soil. Other farmers need to subsoil frequently to avoid severe compaction building up, but it should not be necessary to subsoil more frequently than every third or fourth year.

Direction and Spacing of Subsoiling

If land is pipe drained, subsoiling should be at right angles to the lines of drains so that if water moves laterally in the loosened soil the distance to a drain is minimised. Sloping land should be subsoiled at an angle to reduce the risk both of soil erosion and of wet areas forming at the base of slopes. For maximum benefit, the lifting effect of successive runs should meet, and to achieve this the distance between slits should be roughly equal to the depth of subsoiling as illustrated in fig. 13.6.

With wings the distance apart can be widened to $1\frac{1}{2}$ times working depth. When a twin-bladed implement is used with legs behind the tractor wheels, the closest spacing that can be conveniently achieved is by splitting the work down the middle.

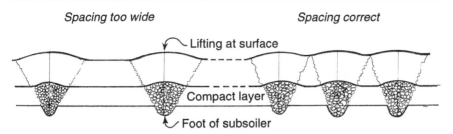

Figure 13.6   Spacing between subsoil runs should be roughly equivalent to depth of working for good shattering

Topsoiling or Shallow Subsoiling

For situations which require loosening to 30 cm or less, conventional subsoilers are much less effective than the implements described on page 161, which lift and crack the soil with minimal rearrangement of the loosened soil. Consequently, the effect is to loosen without raising large clods to the surface except in very dry conditions. Lifting of clods can be reduced further by shallow cultivation ahead of loosening. A drawback of land loosened in this way is the ease of re-compaction.

RISK OF DAMAGE

Deep subsoiling can do more harm than good on undulating land with inadequate natural drainage. Rain falling on to subsoiled land quickly percolates to the depth of working, and unless there is a permeable subsoil, the land soon becomes saturated and the lower areas of the field waterlogged. Consequently, on poorly drained land it is essential that the water moving through the subsoiled layers has access to pipe drains covered with stones. These should be sited in the bottom of depressions and along the base of slopes (fig. 13.7).

In cases where land has been mole drained, and subsequently severe compaction is produced at plough depth, there is a risk that deep subsoiling will disrupt the channels causing deterioration in the drainage. Three solutions are available and the choice will depend on the individual circumstances. It may be decided not to subsoil but to re-pull the moles at 2 m intervals, when the land is dry enough to loosen the compaction. Alternatively, if the moles are in good condition and the compaction does not extend below

about 300 mm, the compaction can be broken with a deep cultivator without any harm to the moles. If the compaction goes
deeper than 300 mm, it should be possible to subsoil in between
the mole runs where these can be located accurately. Recent
experiments have shown that, provided the top of the mole
channel is 7–10 cm deeper than the maximum depth of subsoiling,
the mole channel is unlikely to suffer damage.

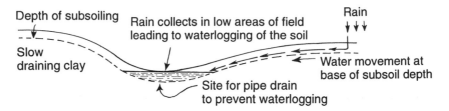

Figure 13.7   Dangers of subsoiling inadequately drained undulating
land

CHAPTER 14

SOIL ORGANIC MATTER

The important but complex role of soil organic matter in crop nutrition, soil structure, and cultivations has been carefully studied both in the field and the laboratory for many years with considerable success. In this chapter we examine the findings of some of this work and discuss the significance of organic matter in different soils.

GROUPS OF ORGANIC MATTER

Three broad groups of organic matter can be identified in soils:

Living Material
A small but very important fraction is made up of a great variety of living organisms. This includes roots of plants, bacteria, fungi, insects, earthworms and soil pests, in addition to many other organisms. This fraction achieves vital biological processes, such as rotting and nitrification, which are essential to life on earth.

Decomposing Residues
After death all the plant and animal residues, including organic manure spread on the land, make up a group of material which is actively decomposing in the soil. This process may release nitrogen if the residue is rich in protein, e.g. clover or cabbage residues, or absorb nitrogen, if the protein content is low, e.g. straw or wood chips.

Residual Organic Matter or Humus
The residual organic matter, or humus, remaining after decom-

position has taken place is a dark-coloured material which gives the characteristic colour to topsoils. This material behaves similarly in most agricultural soils, and although it is much more resistant to breakdown than the residues from which it is formed, it is nevertheless slowly lost by bacterial attack. Most soils also contain some very fine coal and soot-like material.

DETERMINATION OF ORGANIC MATTER

The routine method used for determining soil organic matter in the laboratory estimates the total carbon content of soil. The method cannot distinguish between the three types of organic matter mentioned above, but in arable soils, unless there have been recent heavy additions of manure or other organic material, the carbon figure is due largely to the residual organic matter or humus.

BIOLOGICAL PROCESSES IN SOIL

Even the most lifeless looking topsoil contains millions of bacteria as well as large numbers of other minute organisms in each spadeful. These are responsible for a number of agriculturally vital biological processes, most of which are beneficial, but a few of which are harmful.

The general process of decomposition of plant and animal remains is one in which a large number of different types of soil organisms play a part. In a well aerated fertile loam, remains quickly become unrecognisable to the eye, but in a poorly drained over-cropped soil, straw in particular can be ploughed up nearly unaltered after more than a year. The type of decomposition taking place in the soil depends on the supply of air. If the structure and drainage is good, air enters the soil easily and normal or aerobic decomposition takes place, with the release of carbon dioxide into air spaces in the soil. But if oxygen supply is limited by wet compact conditions, bacteria using nitrates and sulphates as their source of energy, rather than the oxygen in the air, dominate the decomposition process and produce conditions in which plant roots cannot live. These are the foul-smelling anaerobic conditions commonly found in soils after a wet autumn, and identified by grey colours.

Two other important bacterial processes in soils involve

nitrogen. In the first, proteins in plant and animal remains are broken down to ammonium compounds, and these in turn are oxidised by nitrifying bacteria to give nitrates, the form in which most nitrogen is taken up by plant roots. Similarly urea and sulphate of ammonia are transformed to nitrates in the soil. Denitrification, or loss of nitrates from the soil as nitrogen gas, is the second process. Loss of an essential plant nutrient is obviously a harmful process, and denitrification is partly responsible for the extra nitrogen requirement of crops growing in poorly drained land. Denitrifying bacteria only become active in anaerobic conditions.

LEVELS OF SOIL ORGANIC MATTER

In arable land the levels of organic matter in topsoils normally range between 1 per cent and 5 per cent. The most extreme variation is in sands, often with less than 1 per cent, and peats which commonly contain up to 80 per cent. The level in a soil represents the balance between gains from fresh residues and losses from decomposition. The major factors which influence this level are:

Climate
In higher rainfall areas of Britain, and in marshy areas of poorly drained field, the accumulation of organic matter is encouraged by the wetter conditions, so that, for example, an arable loam in Norfolk is likely to have 1–2 per cent less organic matter in the topsoil than a similar arable soil on the coastal plain of Lancashire. When waterlogging is very prolonged or permanent, peat builds up because of the negligible rate of decomposition.

Rotation and Manure
Residues in the form of roots and stubble, returned to the soil by various crops, have been measured by a number of workers. The data in Table 14.1 partly explains the gradual build-up of organic matter that occurs under grass, and the fall associated with the growing of root crops.

In terms of total organic matter in the plough layer before decomposition, all crops, even 3-year leys, produce only small increases. After decomposition these increases are further reduced, and usually analysis is not sensitive enough to pick up the small differences in soil organic matter attributable to one crop. The measured weight of roots produced by winter cereals in the whole soil profile is similar to those under a 1-year ley, but many more of the grass roots are found in the top 200 mm. This difference accounts for the extra benefit to topsoil structure imparted by short-term leys. Another interesting point is that cereals leave several times as much organic matter in the soil as are left by root crops.

Table 14.1   Quantities of organic material returned by roots of various crops

Crop	kg/ha of dry roots in top 200 mm of soil	% increase in total organic matter in top 200 mm of soil before decomposition
1-year grass ley	4500–5500	0.2–0.3
3-year grass ley	6500–9500	0.3–0.5
winter cereals	2500	0.1
spring cereals	1450	less than 0.1
sugar beet	550	less than 0.1
potatoes	280	less than 0.1
red clover	2200	0.1

For comparison 25 t of farmyard manure contains 4500 kg of dry matter which before decomposition increases total soil organic matter in the 0–200 mm layer by 0.2%.

Soil Type

Clay combines chemically with humus, and this gives the organic matter a measure of protection against decomposition. Thus soil texture is an important factor in determining the equilibrium organic matter level. For example, clay loam soils in trials at ADAS Boxworth reached a level of about 3.0 per cent organic matter after twelve years arable, whereas on the light soils at ADAS Gleadthorpe the equivalent level was 1.6 per cent.

Cultivations

Although it was previously thought that cultivation was responsible for loss of soil organic matter and that the more intense the cultivation the greater the loss, recent experiments have clearly demonstrated that this is not the case provided that the amount of crop residues returned to the soil is unaltered.

Liming

High pH encourages oxidation and loss of organic matter, and although the adverse effects of acidity for many crops far outweigh any theoretical advantages of underliming, regular overliming should be avoided, particularly on weakly structured soils.

RATE OF GAIN AND LOSS

The level of organic matter in soil reaches an equilibrium determined by the soil type, climate and cropping system. If the level is initially below or above this equilibrium, then there is a build-up or loss until the equilibrium level is reached. For example, if permanent grassland is ploughed up for arable cropping, there will be a gradual loss of organic matter, which often goes on for more than thirty years, before the new level is established. Annual changes in humus level are small and, for example, the increase due to a 1-year ley, or application of 25 tonne/hectare of dung, will almost always be less than can be measured in the laboratory. Even when as much as 25 tonne/hectare of straw is ploughed in, the immediate increase in organic matter will be less than 1 per cent, and the bulk of this will be lost in decomposition over the next year.

Results in Table 14.2 show the changes in organic matter, measured on the ley-fertility trials at some of the ADAS Research Farms. The increase due to a 3-year ley is small in all cases, and even the 9-year leys have not resulted in large increases. The results

Table 14.2 Effect of leys on level of organic matter in soils from ley fertility experiments

			% organic matter in topsoils			
ADAS Research Farms	Location	Soil	at start of trial	after 3-year ley	after 9-year ley	after 3-year arable following 9-year ley
Boxworth	Cambs	Clay loam	3.1	+0.1	+0.5	+0.0
Bridgets	Hants	Chalk loam	4.4	+0.1	+0.6	+0.1
Gleadthorpe	Notts	Sand	1.6	+0.1	+0.36	+0.3
Rosemaund	Hereford	Silt loam	3.5	+0.2	+1.0	+0.0
High Mowthorpe	Yorks	Chalk loam	3.8	+0.4	+1.0	+0.5

also show that although the build-up under ley is slow, this hard-won increase is quickly lost by returning to an arable sequence. This suggests that much of the build-up is due to partially decomposed material rather than residual organic matter.

ORGANIC MANURES

Many waste products of plant and animal life are used as organic manures, and until the middle of the nineteenth century these materials were the only ways of quickly improving the fertility of land. Since this time simple chemical compounds have assumed an ever-increasing importance in maintaining soil fertility in the UK. However an unnecessary division still persists between advocates of 'organic' farming, who avoid fertilisers, and the majority who rely on fertilisers either partially or entirely to grow good crops. The most reasonable approach is to use whatever organic manures are available, and to supplement these with fertilisers as and when required. Without manufactured fertilisers it would be quite impossible to maintain the high level of agricultural production which has been achieved in the UK.

Quantities Produced

Approximately 50 million tonnes of farmyard manure (FYM) and three million tonnes of poultry droppings—together equivalent to approximately five tonnes for each hectare of crops and grass—are produced annually in Britain. For obvious reasons more of this material is spread on grass than on arable land where the need is greater. On many all-arable farms in southern and eastern England no organic manures are available.

Nutrient Content of Manures

In experiments the nutrient content of organic manures usually accounts for most of the benefits obtained. A dressing of 25 tonne/hectare of well-made FYM provides about 38 kg N, 50 kg P_2O_5, and 94 kg K_2O per hectare, for the first crop, and smaller quantities for the second crop after application. Poultry droppings and deep litter are richer in nitrogen and phosphate than FYM, and the average amount of nutrients available to crops from various materials is shown in Table 14.3. These materials are however very variable in nutrient content and the figures should only be taken as a rough guide.

Table 14.3 Approximate nutrients available from poultry manures for first crop after application

	Rate: 1 tonne of material		
	kg N	kg P_2O_5	kg K_2O
Poultry droppings*	12	9	5
Deep litter manure	12	16	10
Broiler litter	15	15	10

* Fresh droppings may scorch roots if applied in large quantities.

Effects of FYM on Structure

G. W. Cooke in his book, *The Control of Soil Fertility*, concludes from the results of several long-term experiments in the UK, that 'the effects of FYM seem to be explicable mainly in terms of extra nutrients supplied'. However, in most of the experiments better soil structure or physical condition was recorded on plots with FYM, and they could be worked earlier and more easily. At the HRI-Wellesbourne, and at ADAS Stockbridge House Horticultural Station in Yorkshire, both on light textured soils, annual dressings of FYM resulted in increased yields of vegetable crops that could not be achieved by fertilisers alone, and at the latter station continuous vegetable cropping was not feasible without dressings of organic manure. At ADAS Rosemaund, on a silty soil, 30 tonne/hectare FYM applied for the potato crop from 1964 to 1966 raised yield by an average of over 5 tonne/hectare, compared with the optimum fertiliser treatment.

Although normal agricultural dressings of 25 tonnes per hectare FYM have negligibly small effects on the level of soil organic matter, the experiments referred to suggest that on some 'difficult soils' benefits from FYM cannot be attributed to nutrients alone. These benefits are usually attributed to structural effects, although continuity of nutrient supply may also be involved. It was mentioned earlier that regular applications of FYM every 1–3 years have been observed to improve working properties of soils. Although these effects are difficult to translate into increased yields in field experiments, in the context of the management of whole farms these advantages are significant, and likely to result in better yields by virtue of better seedbeds and more timely drilling of crops.

When organic manures, straw, or other crop residues are

applied to wet soils late in the year the result can be to reduce yields. This effect has been noted on poorly drained and compacted soils, and is the result of wheeling damage and anaerobic conditions in the ploughed land.

Straw and Soil Organic Matter

Long-term trials at a number of sites measured the effects of ploughing in chopped straw compared with removal or burning the straw in the field. Results showed negligible increases in soil organic matter from returning the straw and no yield advantages. However, most of the soils were already well structured, and these results do not detract from observations that sandy soils and weak structured silts and sandy clay loams benefit from straw incorporation. The extra biological activity resulting from straw decomposition gradually improves the structural stability of the topsoil.

EFFECTS OF LEYS ON PRODUCTIVITY

Sir George Stapledon in his book, *Ley Farming*, published in 1942, makes a strong case for the benefit of leys grazed by animals as a rotational crop on arable farms. At the present time in arable areas, separation of grazing enterprises from arable cropping is the rule rather than the exception, and the concept of ley farming has declined in popularity particularly in eastern England. This decline has taken place partly because of economic pressures, but social factors and practical difficulties of integrating grazing animals on arable farms have also been important.

The effect of different crops on the yield of crops that follow usually falls into one or more of three categories:

- *Nutritional effects* are usually associated with extra-nitrogen residues left in the soil, for example by legumes.
- *Soil structure effects* are usually associated with variation in quality of tilth. For example, late lifted sugar beet often leaves puddled and compact conditions, and spring cereals leave poorer conditions than winter cereals. Effects of this type are temporary, but soil structure may be influenced more permanently by, for example, the change in soil organic matter following grass.
- *Soil-borne pest and disease* levels are greatly dependent on previous cropping, and one of the major roles of leys is as break crops to reduce the incidence of eelworms, eyespot, etc. The

influence of a previous crop on the weed population may also be significant, e.g. winter cereals encourage black grass.

Since *Ley Farming* was published, several long-term experiments have examined the influence of leys on soil properties, and on yields of arable crops which follow them. Nutritional effects of leys were taken into account by applying different levels of nitrogen for arable test crops. Cropping sequences were chosen to reduce incidence of soil-borne disease and pests to insignificant levels, and any difference in crop yield following a ley, not assigned to nutrition, has usually been attributed to structural differences. Table 14.2 gives the increases in soil organic matter due to the 3-year and 9-year leys in some of these trials.

At a majority of the sites, including Rothamsted, nutritional effects explained almost all of the differences in yields due to leys, but at three sites increased yield after leys could not be explained in terms of nutrition. These sites were ICI's Experimental Farm at Jealott's Hill (sandy loam over London clay), and ADAS Rosemaund (silt loam) and Gleadthorpe (sand). At Jealott's Hill the soil structure declined badly under all arable cropping, and the soil was more difficult to cultivate; cereals and kale were the test crops. Potatoes at Rosemaund yielded about 5 tonne/hectare extra after leys at the optimum nitrogen level. On the light sand at Gleadthorpe, extra water holding capacity after leys resulted in increases of 500–900 kg/hectare of barley.

In general, sites where leys did not give structure effects were on well-structured soils, but even on these, differences in visual structure and working properties after leys were noted in some years, and improvements of this type are valuable on commercial farms. Sites responsive to structural effects of leys were on unstable structured soils.

More experimental work is needed to find the influence of leys on 'difficult soils' using, where appropriate, sensitive field vegetables among the test crops. But it is clear that ley enterprises have to be profitable in their own right, and should not rely for their success on the unpredictable benefits that may or may not be realised by arable crops grown after them.

GREEN MANURING

Mustard, leguminous crops, and other crops grown for green manuring are not harvested, but ploughed in prior to the next crop (see page 259). In general, most or all of the value of green manures is from the extra nitrogen released when the plant residues decompose and usually this benefit can be matched by fertiliser nitrogen. However, occasionally benefits are obtained which cannot be attributed to nitrogen and these may be related to better structure and/or water holding capacity. In one such experiment at Woburn sugar beet after ploughed-in trefoil gave 1 t/ha extra sugar. Unfortunately it is not yet possible to predict the situations in which these effects are likely to arise. Green manures, ploughed in the same season as the next crop, can deplete reserves of soil moisture in dry years, and adversely affect the next crop.

EFFECTS OF ORGANIC MATTER ON SOILS

In general, humus makes both sands and clays more loamy in character, and lends stability to unstable structured soils. More than one agriculturally important property of soil is affected by the level of organic matter, and often the effects are complex and not easily separated. The most important properties influenced are:

Nitrogen Supply

The only important natural source of nitrogen in the soil is contained in organic matter. Supplies of nitrogen from the soil are thus closely related to the level of organic matter. In some soils rich in humus, or in residues of legumes, this source of nitrogen can supply the total nitrogen requirement for the next crop. There is no evidence that nitrates from organic matter are any more effective for plants than nitrates from fertiliser sources, except (i) on very light sandy soils, where fertiliser nitrate applied in the seedbed is more prone to loss by leaching in heavy spring storms, and possibly (ii) for winter cereals on heavy land in wet springs; in these years cereals sometimes show severe yellowing before it is possible to go on the land to distribute fertiliser. Where the previous crop has been a legume, nitrogen is available earlier to the cereal, and the deficiency is less likely to arise.

When large amounts of straw, or other bulky organic materials with low protein content are ploughed in, nitrogen deficiency frequently occurs in the following crop, unless extra fertiliser nitrogen is applied. One tonne of straw requires 20 kg of nitrogen for decomposition.

Stability of Structure

Organic matter helps to prevent breakdown of structure by water. This role is important on soils with naturally unstable structures (see Chapter 6), and on sandy soils subject to wind erosion. An extra 1–2 per cent organic matter is required before surface capping is materially reduced in susceptible soils.

Water-holding Capacity

Higher levels of organic matter are associated with greater water-holding capacity, and on very light soils this can reduce risk of drought appreciably. This effect has been recorded on sands at ADAS Gleadthorpe, where cereal yields are higher on land after 3-year leys because of higher available water content.

Overcompaction

A soil with a high organic matter content will normally be more open and less easily over-compacted than one of lower organic content. The results in Table 14.4 show this effect for a light soil with contrasting levels of organic matter in the same field. For this effect to be agriculturally significant a difference of $\frac{1}{2}$ per cent or more in humus level is needed.

At the other extreme, in soils where the organic matter content is above 10 per cent, puffiness becomes a problem particularly in dry springs.

Table 14.4 Effect of extra soil organic matter in resisting compaction of a silty soil

% organic matter	Position	Soil density, g/cc	% pore space
2.9	In wheelings	1.49	45
2.9	Out of wheelings	1.33	50
5.3	In wheelings	1.25	55
5.3	Out of wheelings	1.16	59

Plasticity and Cohesion

The workability of heavier soils is related to their plasticity or putty-like behaviour when wet, and their tendency to form strong clods when dry. Both of these tendencies make cultivations more difficult. This situation is most likely to occur when the organic matter level is low; higher humus content masks plasticity, reduces clod problems and encourages friability. Differences are readily noticed on clay and silt soils, where the difficulty of cultivating old arable fields contrasts with the ease of managing fields more recently ploughed from grass.

SIGNIFICANCE OF ORGANIC MATTER IN ARABLE SOILS

Every farmer knows the excellent soil conditions that persist for many years after old pasture is ploughed out. However, the experimental results already discussed clearly show that it is unrealistic, and indeed virtually impossible, to attempt to recapture these conditions within a predominantly arable system of long standing. In many soils of eastern and south-eastern England organic matter levels are very low, because of the long tradition of arable farming and the relatively dry climate which encourages humus loss. For the naturally well structured and well drained soils, for example, chalk and limestone soils and many medium textured loams, the level of organic matter is not critical. On these soils the small increases in humus that can be achieved with leys give only marginal advantages in soil management, which are unlikely to benefit most crops.

However, other soils are more sensitive to the level of humus and on these quite small increases can give greater latitude in soil management. Soils which fall into this category are those with unstable structure and/or high clay content. In general, the higher the annual rainfall, the more likely these soils are to benefit from leys in arable systems. Unstable soils, for example, the Lincolnshire and Norfolk silts and sandy boulder clays of Suffolk, undoubtedly benefit from ley farming, but this does not ensure any financial advantage. Indeed unless the grass enterprise gives high returns in its own right, then the returns are likely to be lower. On many of the traditional grassland clay areas of eastern England and the East Midlands (for example, parts of Northamptonshire, Leicestershire and Huntingdonshire), and on London Clays in Essex, Hertfordshire and Buckinghamshire, arable systems are very vulnerable to wet seasons, and grass with arable, or grass-only systems are better suited for maintaining good soil structure. However, adequate field drainage and faster cultivation systems have in part reduced the risk attached to arable systems in these areas. Advantages associated with grass can be due to greater drying of soil, and lack of damaging cultivations, in addition to the nutritional, structural, and disease effects already noted.

We have seen that the level of organic matter influences soil behaviour in a complex manner, which varies in importance from soil to soil. In general, higher levels of organic matter mean greater flexibility for timing of cultivations, and of other aspects of soil management; in other words, it provides a buffer against mis-management. The individual farmer's skill as a cultivator is a vital factor in trying to assess the importance of the humus level on a particular soil. Some can overcome problems which to less skilful operators would be overwhelming, and it is for this reason that trying to establish critical levels of organic matter as guides to the management of particular soil types can be misleading and tends to divert attention from the real job in hand, that of developing techniques of cultivation to suit the soil conditions that prevail.

CHAPTER 15

CULTIVATION SYSTEMS FOR CROP AND SOIL

During the last three decades chemical control of weeds, decreasing labour, increased mechanisation, and demand for higher quality in many crops, have prompted fundamental rethinking of traditional systems of cultivation. While there are sound reasons for resisting precipitous change from well-tried systems, there are also compelling reasons for arable farmers to experiment on a limited scale with new approaches and new implements, and in this way to gradually evolve modified systems which fit their individual requirements better. In this chapter we examine the aims of managing clays, sands, silts and peats and then discuss soil management for potatoes and small-seeded crops.

SOIL EXAMINATION FOR BETTER MANAGEMENT

All cultivation systems, but in particular shallow cultivation, are only successful where farmers are able to identify compaction reliably and thus take action before yields are reduced. Soil examination is the key aid to cultivation management and its value has already been referred to on pages 85 and 190.

Topsoil Examination
Topsoil examination is mainly concerned with decisions on depth and type of cultivation on medium loam and clays. Fields should be examined when moist. March and April are ideal months, whereas the soil after harvest is often too dry and this makes assessment unreliable. When examining soil structure in the field, the aim is to decide **whether the soil condition is adequate for water movement and root penetration.** This must be assessed with

a range of weather in mind—poor soil structure is most likely to limit yields in abnormally wet or dry seasons. The top 300 mm is exposed by taking a spadeful of soil; if the soil is compact, it is often necessary to take a second spadeful after opening the hole before attempting examination. To assess a field it is generally necessary to look at the soil in ten to fifteen places.

If the soil is moist the ease with which the spade penetrates is a useful indication of soil condition. Once removed, the spadeful of soil is examined. The first criterion is how porous the soil appears. It may be loose and friable or contain large dense smooth-faced clods fitting closely and containing few pores. If there is a crop in the field, root development will indicate how well the roots have been able to explore the soil. Examination should include gentle breaking up of the soil along any lines of natural weakness.

When examining the soil, the depth of previous cultivation is often apparent. It is likely that any poor soil structure will be found just below this depth, especially if the field has been cultivated at the same depth for several years. In all fields there will be variation due to change in soil texture and structural condition—old field boundaries, changes in slope, old wheelings, wetter areas and headlands are all sources of potential variation. Examination should include problem areas which may require different treatment from the majority of the field—often these areas will dictate the overall field cultivation.

Interpretation

Colour photo 12 shows a well structured clay topsoil which is ideally suited for shallow 6–8 cm disc or tine cultivation if the straw is baled. Deeper tillage will bring up clods which would lead to coarser drier seedbeds for autumn-sown crops.

Colour photo 13 is a clay topsoil in which oilseed rape rooted only poorly. Prior to the next crop, loosening is required throughout the topsoil (0–25 cm). If the next crop is winter wheat the best way of achieving this would be to loosen the surface with heavy discs followed by the type of shallow profile looseners described on page 161. These 'shallow subsoilers' crack the soil while leaving the surface comparatively free of large clods. If the next crop is spring sown, ploughing when the soil moistens would be the best option.

Examination of Subsoils

To assess the worthwhileness of subsoiling, examination needs to be carried out on the 300–450 mm layer in four or five pits in a field. A spade or tractor-mounted digger should be used. On light soils, compaction is best diagnosed by finding resistance to the spade when digging, followed by picking the exposed soil face with a knife or trowel. Beware of differences in resistance due to moisture; ideally the profile should be moist to depth. If there is an obvious compact layer with looser soil above and below, loosening is necessary. On heavy soils subsoil pans are rarely intact, usually being broken by cracks and fissures. Where the subsoil is composed of large dense structures subsoiling may well be beneficial to rearrange and open up more cracks in this layer. In wet conditions water may be seen to be held over a plough pan and it is often possible to find horizontal rooting. Examples of compact subsoil are illustrated on pages 82 and 83 and colour photo 15.

MANAGEMENT OF SANDS

The major limitation of sands is water shortage caused by low water-holding capacity and the key aim is to select cropping and growing systems which minimise losses of yield due to moisture shortage. Lower clay and organic matter contents accentuate this risk (photos 15.1 and 15.2). In south-east and eastern England, where rainfall is low and transpiration high, drought risk is much greater than in areas further north and west. For example, the number of years in twenty when a soil moisture deficit of over

Plate 15.1   Growth patterns in a Norfolk sugar beet field caused by variation in severity of drought on soils of different texture
(Crown copyright)

Plate 15.2   Polygon and stripe patterns in sugar beet growing in Norfolk. Regular variation in depth of shallow glacial sand over chalk is responsible for this drought pattern (Crown copyright)

75 mm occurs on the Breckland sands near Cambridge is 15, whereas on the sands of the Lancashire coastal plain it is only 6 in 20. The map in fig. 15.1 gives similar data for England and Wales.

For farmers on sands in drier areas four strategies are useful in minimising effects of drought stress:

- Selection of crops which are less susceptible to drought such as winter barley, sugar beet and carrots, rather than more susceptible species such as potatoes and winter wheat.
- Early establishment of both autumn- and spring-sown crops, so that as much growth as possible is made during September-April, a period of relatively plentiful moisture supply.
- Choice of rotations which minimise the risks of root damage from nematodes and other soil pests and diseases.
- Encouragement of deep rooting to exploit fully the reserves of stored water. This can be achieved by avoiding both underconsolidation in the seedbed and compaction in the profile.

Nutritional Disorders

Inability to hold good reserves of nutrients—other than phosphate—and ease of leaching, make sands susceptible to numerous nutritional problems, of which acidity and nitrogen deficiency are the most damaging. Nitrogen deficiency is enhanced by the ease with which nitrates are lost by leaching, but in addition the low

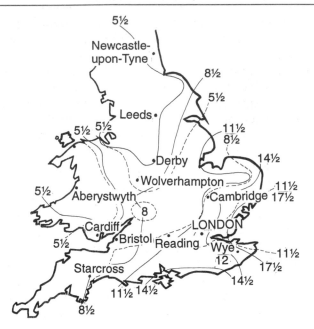

Figure 15.1   Number of years in 20 with soil moisture deficit greater than 75 mm

level of organic matter in sands leads to a relatively low supply of nitrogen from the annual decomposition of organic matter. Nitrogen, phosphate, and potash deficiency can readily be made good with conventional fertilisers, but deficiencies of magnesium, manganese and boron are also widespread and these need special treatment (see Chapter 3).

Cultivation

Aggregation in sands is at best weakly developed and in consequence cultivation tends to lead to extremes of density. At the one extreme it is difficult to avoid loose underconsolidated conditions after ploughing, and at the other extreme the wheelings of traffic used to consolidate the seedbed and spread fertilisers cause tight compact conditions. Not all sands compact to densities which restrict rooting and it tends to be those with a wider range of particle sizes which are more likely to pack tightly as smaller particles fit closely between the larger ones under the influence of applied forces. Driven wheels are a major source of this compaction but due to the weak structure even surface tension exerted by drying is sufficient to cause significant increase in density.

Experience and experiment have both confirmed the common-sense view that zero and shallow cultivation systems are likely to lead to damaging compaction on many sands and are therefore unsuitable techniques for these soils. The mixed arable rotations common to our sandlands, and their ease of working, make the mouldboard plough the first choice for primary cultivation. The difficulty of producing firm seedbeds after ploughing have largely been resolved by coupling a packing roll (furrow press) to the plough so that a firm level seedbed may be produced with at most one further pass with harrow and levelling board. On the lighter sands this technique ensures that firm seedbeds can be produced immediately before drilling both in autumn and spring without the need for a lengthy period of natural settling. To minimise drought effects in later stages of growth the optimum period for establishing winter cereals is mid September to early October, and December to the end of February for spring cereals. Phosphate, potassium and magnesium fertilisers should be applied prior to ploughing to avoid wheeling damage on cultivated land. When winter frost is prolonged previously firm seedbeds can become puffy again and it is interesting to note that in Scotland spring cereals sown into loose seedbeds have been shown to respond well to combine drilled fertiliser.

In exposed landscapes, for example the Vale of York, sands are very susceptible to wind erosion and need special techniques to avoid damaging soil and crop losses (see Chapter 17).

When sands are ploughed at about the same depth every year pans develop below this depth and are not disrupted as the soils dry. Irrigated sands are particularly prone to this form of panning because they tend to be ploughed in moister conditions. Regular close-spaced subsoiling (see Chapter 13) is needed to loosen this layer, or alternatively short tines attached to the plough bodies will probably effect the required loosening.

Level of Organic Matter

Organic matter levels in arable sands are usually very low except on old heathland. An extra 1 per cent gives better water retention and resistance against overcompaction (see page 205). However, in an arable system it is impossible to achieve this increase without long-term leys or very large amounts of farm manure which are rarely available. If leys are grown on light land, they should be part of a profitable livestock or grass seed enterprise, and any benefits for following crops considered as a bonus. Unfortunately grass is even more susceptible to drought than many arable crops, and in dry areas production is badly restricted in most years. Crop yields from sands benefit more from regular and generous application of animal manures than those of other soils.

MANAGEMENT OF CLAYS

Limitations
Clays are here defined as soils with 30 per cent or more clay-sized particles in the topsoil. Usually clay subsoils contain over 40 per cent clay and as much as 80 per cent clay in some alluvial situations. Although this definition encompasses a wide range of soils and soil properties the 'clays' have these common properties:

- The soil matrix swells and shrinks as it wets and dries. This encourages self-structuring and surface tilth.
- When the subsoils are fully swollen by winter rainfall permeability to water is low or very low.
- The soil moisture range for satisfactory cultivation is small; this friable range is bounded by very hard consistency on the dry side and plastic conditions on the wet side.
- Surface evaporation is needed to bring a wet but drained clay soil into a moisture range suitable for cultivation.
- When wet, clays are particularly susceptible to damage from wheels which compact and smear thus reducing water permeability to unacceptably low levels.
- Crops on clayland suffer less from moisture shortage than those on many other soils. Potential yield is therefore very high provided the other limitations can be circumvented.

An optimum management strategy for clays can be derived from the common properties listed, which in all but very abnormal economic situations will give optimum returns. All clays require artificial drainage including mole drains. Crops which have to be harvested from within the ground—root crops—should not be grown, and crops established in the autumn when the soil is relatively dry should be grown in preference to spring crops.

These requirements indicate that grass and autumn-established combinable crops are the most favoured: grass because regular cultivation is avoided and because it imposes long periods of drying on the soil; winter crops because they are established during the most favourable period for cultivation and harvested during the period of near-to-maximum soil moisture deficit. In practice, grass tends to be the prevalent crop of the wetter claylands, and autumn cereals, winter oilseed rape and winter beans the dominant crops of the drier clayland areas.

Drainage and Work Days

In arable systems the control of drainage water is essential on clay land, but because of slow water movement in most clay subsoils, pipe drainage alone is insufficient, and a permeability treatment is essential. Mole drainage or subsoiling, if they are carried out at the correct moisture content (see page 67), increase the permeability sufficiently to prevent water building up in the subsoil, but it is essential to have permeable backfill over the drains. In general, provided pockets of lighter soil are not too common, mole drainage is more effective than subsoiling on heavy land.

Unfortunately even when all drainage water is removed, clay soils are still too wet for cultivations, and evaporation of water from the surface must take place before cultivations can start. This limitation is illustrated in Table 15.1, where the number of work days in the spring is compared for heavy, medium, and light land in the East Midlands. A work day is defined as a day on which field cultivations would be satisfactory.

Table 15.1 Influence of rainfall and soil texture on the average number of work days

| | Number of work days | | | | | | | | |
| | February | | | March | | | April | | |
	Light	Medium	Heavy	Light	Medium	Heavy	Light	Medium	Heavy
Wetter than average	3	2	0	16	14	9	21	19	16
Average rainfall	8	5	3	25	24	20	26	23	16
Drier than average	11	9	8	29	29	27	28	26	25

Data from C. V. Smith, Meteorological Office, Bracknell—Agricultural Memo. 438.

Compaction and Cure

The risk of damage to structure on heavy land is high, both because of the shorter number of optimum work days, and also because of the ease with which the topsoil is smeared and puddled. This type of damage is an important cause of waterlogging and yield loss on clay soils, and must be reduced to a minimum if high yields are to be achieved. Tramlines are a major aid in maintaining good physical conditions.

The annual drying cycle within the soil profile is very important for the reconditioning of structure. After dry summers and autumns, heavy land is invariably improved and drainage functions well. The transpiration of a vigorously growing crop is the key to this drying process, but less drying takes place where structural damage restricts rooting, and this is often the initial cause of gradually worsening soil conditions. It is mainly in this context of effective drying cycles that grass leys play such an important part on clay soils. In addition, where half or more of the rotation is ley, the organic matter status of the land will be $\frac{1}{2}$–1 per cent higher, and this increases the nitrogen supply from the soil, and improves the fertility and working properties of the land (see Chapter 14).

Cultivation

Clays are the most difficult soils to work and cultivation should be avoided or minimised unless there is a clear need. A major aid in determining cultivation need is soil examination (see page 208). For most clays very shallow cultivation to encourage weed germination after straw removal, coupled with deeper topsoil loosening every 3–4 years and mole drainage every 4–7 years will ensure soil conditions that do not limit yield (see Chapter 16). The deeper loosening of the topsoil should be achieved without bringing up large clods; this can be done in two ways. The first method is to loosen in stages, shallowly at first with pig tines or discs and then deeper with heavy rigid tines to below the depth of compaction. This method loosens effectively and incorporates chopped straw well. The second method is to use an underloosening implement designed to loosen with minimum surface disturbance (see page 161 and Plate 10.10). This technique gives less loosening than the first but leaves a better surface condition for crop establishment. Loosened soil is very susceptible to recompaction and whichever method is used drilling should be, wherever possible, the next operation. Of the winter crops, oilseed rape is most susceptible both to topsoil compaction and to coarse dry tilths. Consequently loosening, if required, should be done in the autumn preceding the oilseed rape crop.

Spring-sown crops normally benefit from deeper autumn cultivations. Ploughed or deep cultivated land left over winter to break down, helped by a well-timed levelling pass on a frosty morning, makes a very satisfactory primary cultivation. The key to spring work on clays is one pass only without curved tine implements. The latter are designed to bring up clods, which then have to be broken down by further cultivation, each pass causing more wheelings on the wet clay.

The poorer and less well managed clays tend to have coarser-structured subsoils and often tighter structure in the lower topsoil. Where these physical conditions are present a programme of progressively deeper loosening over a period of years will improve the potential and eventually lead to a soil which can be shallow cultivated.

MANAGEMENT OF SILT, WARP AND BRICKEARTH SOILS

Silty soils fall into the following textural categories: very fine sandy loam, silt loam, sandy silty loam and silty clay loam. They are dominated by particles in the size range 0.002–0.1 mm which feel smooth and silky to the fingers. Soils of this type are found in the Trent and Humber warp lands, in the East Anglian Fens, in the brickearth of Sussex, Kent, Essex and Norfolk and some of the red soils of Hereford and Worcestershire.

Potential for Crop Production
Deep silty soils have the highest water-holding capacity of any mineral soil. Crops are only affected by drought in dry summers when yields of shallow-rooting crops such as potatoes and onions are limited. This characteristic is very important as many of the areas of these silty soils occur in the driest parts of England. Fenland silts have water tables at $1\frac{1}{2}$–3 m depth which benefit crops by water movement upwards as soil above is dried out.

The lighter silts (very fine sandy loams) are ideal for a very wide range of crops and frequently grow vegetables and root crops besides cereals. The heavier silts (silty clay loams) are more difficult to work and less suited for vegetables. Most of the Grade I land of the UK has a deep silty profile.

The main problem of silty soils is their poor stability to water both in the form of waterlogging and rain impact. Their lack of clay and medium-coarse sand causes them to slump and cap to a closely-packed state containing very little air. Efficient field and arterial drainage is necessary wherever these soils are located in lowland basins.

Soil Structural Problems
Soil organic matter is very important in determining the working

properties of silty soils; it is the main constituent contributing to stability. Over a century of mixed arable farming, organic matter levels fall to an equilibrium level of 1.8–2.5 per cent; there is no specific danger level for a soil but the lower the content the more care has to be taken with management.

Soil structure problems in silts can be dealt with in three areas:

- At the surface—capping
- Within the topsoil
- In subsoil

Except where severe, capping is only of agricultural importance with small-seeded crops which are prevented from emerging; for example sugar beet, onions, lettuce and carrots. Although the effects of capping can be reduced, the risk of poor emergence is always present and this may restrict the area of sugar beet and vegetable crops which are drilled to a stand. The impact of rain falling on to unstable aggregates is the cause of capping, and the process is encouraged by a fine consolidated surface tilth. Under these conditions heavy rain seals the surface, water builds up and the partially dispersed soil become suspended. After rain ceases the particles separate out in oriented, interlocked layers which can be peeled apart on drying. Except for small seeded crops, the caps formed on very fine sandy loams are rarely enough to cause reductions in emergence, but on silty and silty clay loams the extra rigidity in the dry cap due to the higher clay content presents a more serious obstacle for emerging seedlings. The worst conditions are caused by heavy rain before emergence followed by prolonged dry periods. If the cap remains damp then emergence will be little affected, but if there is a prolonged wet period anaerobism within or below the cap can damage roots of seedlings and produce uneven growth. Care should be taken not to make seedbeds that are too fine. Use of polymer conditioners may be worthwhile to give surface stability for high value crops.

The wheels of tractors, harvesters and laden trailers are the major source of topsoil compaction and puddling, but the compacting effects of implements themselves, particularly discs and spring tine harrows, are also significant. Temporary anaerobic conditions are often associated with compaction both in the zone of high density where macro-pore space and permeability have been reduced, and also in the looser overlying material which becomes saturated in wet periods. The condition is identified by a duller colour in the soil, which becomes grey and foul smelling in extreme cases. Anaerobism occurs frequently following the lifting of sugar beet in a wet time, when the resultant mixture of readily decomposable beet tops and puddled soil is ploughed under. Complete burial takes place more easily on the lighter soils, and it is on these that the classical anaerobic layer is most often seen.

The actual damage caused by compaction in a surface horizon is related to subsoil quality. In most of the silty soils of this country, subsoil conditions are excellent and, where crop roots are eventually able to penetrate topsoil compaction, there is an opportunity for compensation in crops with long growing seasons. Where depth of rooting in the subsoil is restricted for any reason, compaction in the surface is likely to be more damaging.

Although highly porous, the weak structure of wet, low-strength subsoils in silty soils predisposes them to the formation of dense layers, usually extending from 50 to 200 mm below the topsoil and referred to as plough pans. The precise cause of these pans is often difficult to determine, but compaction in the furrow bottom, compaction below plough shares working at one depth over several years, and deep compaction under surface traffic, seem to be the factors responsible. Where water builds up over mechanically compacted and smeared layers, slaking of aggregates can take place and the closely packed particles lie on top of, and add to, the thickness of the existing pan.

In cases where plough pans are severe enough to stop root development completely, drought symptoms are likely to occur in any crop, but frequently root growth is only retarded and whether a crop will be affected depends on weather and crop species. On heavier soils, pans crack as roots extract water from them and so complete cessation of root development is less likely on these than on lighter textured soils. Wilting of sugar beet on land with subsoil pans is a useful diagnostic tool for farmers and advisers. In soil with compaction at several depths rather than at one depth, crops will be proportionately worse affected.

Cultivation

Ploughing is the normal form of primary tillage for silts. The intensively cropped light silts are often ploughed three times for three crops in two years. Rotation on these soils often demands that ploughing is adopted to bury the trash from the previous crop and also to remedy soil structural damage in the topsoil. The natural tendency of light silts to slump and the common harvesting of crops under high soil moisture conditions makes the latter aspect very important.

In favourable situations where the previous crop was harvested under good conditions, tine cultivation is adequate prior to drilling of winter wheat as long as the residual herbicide used permits non-ploughing. The rotation is normally less intensive on these soils. Shallow tillage is not normally recommended on light or medium silts due mainly to the lack of topsoil stability and consequent risk or surface ponding and compaction in the top 150 mm.

Seedbed cultivations are critical to the successful establishment of many crops. For small-seeded, precision-sown crops the reduction of wheelings to a minimum is important. The seedbed should not be too fine or the risk of capping is increased. To extend the rotation and grow sensitive crops on the heavier silts a bed system may be adopted but is only occasionally found in practice. Artificial anticapping agents to prevent capping are now commercially available.

Subsoiling is commonly necessary on the light silts and also on silt loams over light silt. On many farms it is carried out routinely to reduce the effect of the inevitable plough pan to a minimum. On deeper silt loams subsoiling may be required but should only be carried out where subsoil water movement is satisfactory or an effective drainage system is present.

MANAGEMENT OF PEATS

The largest areas of peat in arable cropping are in Lancashire and the Fens. Although peats are highly fertile, they do present certain management problems which are not often found in other soils. From the time a peat is first drained, continuous wastage by oxidation is inevitable; this loss usually proceeds at the rate of about 25 mm each year. During the time that the depth of peat is greater than 1 m, and the organic content of the profile high (50–90 per cent), wind erosion is a serious threat to spring crops, and in particular to sugar beet, carrots and onions, which require fine seedbeds and do not cover the ground for several weeks after drilling.

Lime and Nutrients
Acid moss peats of Lancashire and the Fens need a very large quantity of lime before even tolerant crops can be grown. In the Fens some of the most acid peats are described as drummy; sugar-beet waste lime applied at rates in excess of 120 tonne/ hectare is found to be the best liming material.

Nitrogen supply is often excessive on peats, and this is thought to be a factor in the rather poor cereal yields on this land. Sugar percentage in beet is also low for the same reason. Trace element deficiencies in the form of manganese and copper are common on calcareous and overlimed acid peats. However, probably the biggest scourge of peat farmers are weeds, which grow in great profusion and are difficult to control.

Problems of Wastage

Eventually when the peat has nearly gone, the nature of the underlying mineral floor influences the fertility of the land. Commonly a layer of extremely acid drummy peat occurs immediately over the sand or clay, and frequently this stops roots from growing any deeper. In dry years these layers cause patchy low-yielding crops. In recent years deep cultivations to a depth of 750–900 mm have been used successfully to overcome this problem, and to aid the transition from a predominantly organic soil to a mineral soil. Prior to mixing, the topsoil is heavily limed with sugar-beet factory lime. This is then mixed with the acid subsoil either with the aid of a very large single-furrow mouldboard plough ($\frac{1}{2}$–1 m deep) or a double digger plough which has two mouldboards, the second ploughing in the furrow opened by the first.

CULTIVATIONS FOR POTATOES ON MEDIUM AND HEAVY LAND

Where potatoes are grown and mechanically lifted the aims are:

- To grow a maximum yield of good shaped undamaged ware potatoes.
- To produce a ridge which can be lifted quickly, with as little hand labour as possible and,
- To control weeds adequately.

The method of cultivation plays a vital part in achieving these aims.

The first requirement is for good drainage to prevent any risk of flooding in the ridges, and to make possible timely cultivations. This may involve subsoiling the previous autumn. Ploughing should be finished early in the autumn, before the soil is too wet, and should be as level as possible so that frost tilth is not irretrievably lost into deep hollows. Many farmers use clod separators in the autumn to help prepare clod-free ridges in spring. During wet springs this practice leads to slumping of ridges in weak structured soil.

Ridge Size and Shape

Fig. 15.2 summarises the ideal size and shape of freshly made ridges for 750–900 mm rows. Experimental work indicates that a cross sectional area of about 750 sq cm of mould is the ideal size. This is big enough to contain a 50 tonne/hectare crop without undue greening. A ridge of this size has a 200 mm depth of mould, above the uncultivated soil, and a 550–650 mm base when freshly made.

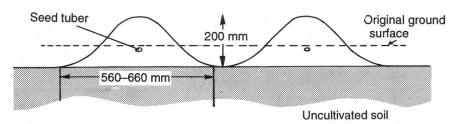

Figure 15.2   Ideal potato ridges

Seedbed Cultivations

There is no 'best method' of preparing a tilth for potatoes which can be used in all circumstances, but the most satisfactory employ straight tine cultivating implements, and not curved tined harrows, which tend to peel up clods (fig. 15.3).

Only 75 mm of tilth is theoretically needed to make the 750 sq cm ridge for a 750 mm row. However, in practice planters need about 125 mm of tilth for smooth working, and this depth allows for some

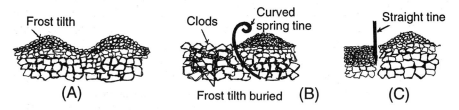

Figure 15.3   Influence of tine shape on final seedbed tilth: (a) ploughed land in spring; (b) clods brought up by spring-tine cultivator and frost tilth buried; (c) frost tilth left on surface by straight tine harrow

inevitable loss of mould during ridge building.

The surest way of producing clods on heavy land is to cultivate too early before the soil has dried sufficiently. Once the land is dry enough, Dutch harrows with levelling board and crumbler roll or powered rotary harrows and spiked rotavators are used to break down further the weathered mould. The number of passes depends on soil type and condition and type of implement, but usually two or three will give sufficient depth of tilth if there has been enough frost. Every extra pass involves more wheelings, loss of tilth already made and formation of clods. After mild winters there may not be sufficient weathering on heavy land, and it will then be necessary to make extra tilth, after planting, with an inter-row cultivator.

If the tilth is too fine, it tends to run together in the ridges, particularly on unstable structured land, but if it is too coarse, mechanical lifting is slowed down, and in dry autumns potato damage is greater. The ideal tilth should contain as few hard clods larger than 40 mm as possible—that is, clods which will be elevated by the harvester—but a fairly high proportion of smaller clods to keep the ridge from slumping in wet weather. If sufficient tilth is present at planting, many growers put up the finished ridge at this time and use a residual herbicide for weed control. In this way clods formed by wheelings are minimised and moisture is not lost during remaking of the ridge.

Row Width and Wheelings

Before planting, every fresh set of wheeling compacts more tilth, forms more clods and, equally important, increases the level of compaction in the layer of soil which will eventually lie immediately below the ridge and through which roots must grow. Three practices are available for growers on heavy soils to offset these problems. The first method is to ensure that all passes from ploughing onwards, including fertiliser spreading, take place in the same wheelings. To accomplish this, ploughed land should be accurately marked out in the autumn with a tractor, and markers used at each cultivation. In this way compaction is restricted to the minimum area possible, and soil under the eventual ridges is untouched by wheels. The second method is to protect tilth that would have been compacted under wheels during cultivations, by moving it to the side with ridging bodies mounted in front of the tractor. Lastly, there are definite advantages associated with ridges wider than 750 mm. If the same 750 sq cm ridge is put up at 900 mm centres, the tilth required is more readily available. Another important advantage of wider ridges is the greater space available for wheelings between the ridges (fig. 15.4). This increased space helps to reduce the compaction, which takes place at the edge of the ridges in narrower rows. 900 mm rows allow the crop to be harvested faster, on average the saving in time is 17 per cent.

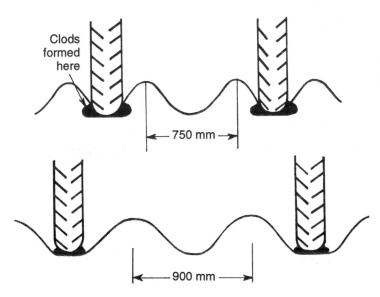

Figure 15.4   Wider rows for potatoes reduce the number of clods produced at the edge of ridges

SUGAR BEET AND SMALL-SEEDED VEGETABLES ON LIGHTER LAND

Seedbed Preparation

Small-seeded crops are very susceptible to establishment problems and have important seedbed requirements during early stages of growth. The seed must be in close contact with moist soil for germination to take place, and the soil beneath the seed should be loose enough to allow rapid root extension when the temperature permits. If roots are unable to grow quickly and to take up water and nutrients, the small seed reserves are rapidly depleted, and the seedling is open to attack from soil pests and diseases. With tap-rooted crops like sugar beet and carrots even slight compaction can cause damage to the root tip, resulting in fanged roots at lifting time.

In the ideal seedbed these requirements are met by having reasonably fine and sufficiently consolidated tilth in the top 25–50 mm, to ensure good contact between soil and seed, and to conserve moisture in the seedbed. Below this layer the tilth should be loose to allow rapid root growth. The tilth should not be too fine, otherwise there is a risk of emergence problems due to surface capping. Not infrequently, efforts to achieve a uniform, fine and level surface tilth are at the expense of layers below, which become over-compacted by wheelings. The seedbed then consists of 25–50 mm of fine tilth, with a compact layer below which prevents the seedling growing away quickly (fig. 15.5).

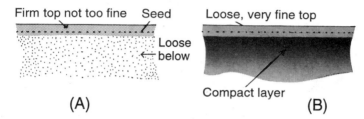

Figure 15.5   (a) An ideal seedbed for small-seeded crops. (b) A poor seedbed for small-seeded crops

The ideal seedbed starts with early level ploughing, so that the minimum of work is needed in the spring. If light soil is ploughed late and wet, a compact layer is likely to persist over winter below the frost tilth, and will be further compacted during seedbed preparation.

Choice of Implements

Where autumn cultivations have left an even loose plough layer, the best seedbeds are prepared with the fewest cultivations. Straight tine implements give the best results because they leave a fine tilth at the surface, and usually only one or two passes are needed. Crawlers give less compaction than wheeled tractors, but where the latter are used cage wheels should be fitted, and the tractors need to be well loaded to reduce slip.

Finger carrots and parsnips for prepacking are usually grown on very light easily panned land and need at least 200 mm of loose soil to give regular shaped, unfanged roots. A deep cultivation at about 250 mm with a strong Dutch harrow plus levelling board gives an excellent tilth to drill these crops into, provided control of

Plate 15.3   Poor emergence in wheelings from cultivations at right angles to the direction of drilling   (Crown copyright)

drilling depth is good enough. If the drill press wheel does not give a firm enough surface, the drill can be followed by a light roll, unless capping or wind erosion are anticipated.

Bed Systems

During the period from ploughing to drilling, fields are criss-crossed with numerous wheelings which may cover more than 90 per cent of the field. These give rise to uneven consolidation, patchy plant population (photo 15.3), uneven growth rate, and in vegetables variable quality of harvested produce. The bed system has been devised to divorce crop rows from wheelings completely, and the crop is grown on land that has had no wheels over it since autumn ploughing (photo 15.4). This is achieved by deliberately marking out the beds with a tractor as accurately as possible, and then using these wheelings for all subsequent operations. It is important that the width of all implements used fits the basic layout, so that no fresh wheelings are needed. Compaction in the wheeling gaps would adversely affect growth of plants on either side, and to compensate for this an extra 150–200 mm is needed between the two rows (fig. 15.6). For example, for carrots grown

Plate 15.4 Bed systems avoid the wheeling problems seen in photo 15.3. All wheelings are confined to uncropped land
(Crown copyright)

on 350 mm rows, the wheeling gaps should be 500–550 mm wide. This technique is not difficult to work out in practice, and could be used much more widely on soils of all textures.

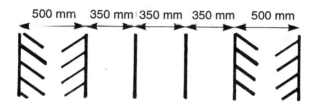

Figure 15.6 Bed system suitable for vegetable crops with crop rows indicated by vertical lines

CHAPTER 16

MINIMAL CULTIVATION

It has long been known that in well-structured soils with adequate drainage and no impediment to root development a shallow seedbed is all that is required for cereal growth. This principle was developed, notably at ADAS Drayton, into a successful system of shallow cultivations, based mostly on tined implements, and elsewhere into zero tillage or direct drilling. Most weed problems that occur with non-mouldboard ploughing and direct drilling can be dealt with by present-day herbicides and there do not appear to be major disease problems.

The chief economic advantage to the farmer of the new cultivation systems lies in maximising winter wheat planting. The yield advantage of winter over spring sown cereals might be of the order of 1 tonne/ha. It has always been possible to equip for entire winter wheat planting on a cereal farm, but the result is likely to be a peak labour requirement in the autumn that is not matched at any other period of the year, and a massive investment in machinery for that peak period. New cultivation systems have streamlined the autumn planting period, so that labour and machinery costs have been contained, while work rates have improved.

Another important use of minimal tillage is in the establishment of oilseed rape quickly and with little water loss in late August. Some of the techniques used in reduced cultivation demand less of the operator than the traditional systems. For example, tined cultivation requires considerably less skill than mouldboard ploughing, but generally the management skill required is greater and a greater knowledge and understanding of soil problems is necessary under the low input systems.

The economics of reduced cultivation and direct drilling depend on a variety of factors. Important among these are:

- It is generally not economically sensible, or technically necessary, to accept any loss of yield as a result of the newer systems.
- The gross return advantage of winter over spring cereals is considerable.
- A large part of the cost of farm machinery is in ownership (depreciation, interest on capital, insurance and housing) rather than operation. In spite of this, it has over many years been profitable to retain the traditional machinery complement, and even add to it with specialised minimal cultivation or straw incorporation equipment, when extra winter wheat planting is involved. Lower returns from cereals may change this situation.
- Differences in chemical cost between the various systems are generally not important. Sterile brome (*Bromus sterilis*) and blackgrass (*Alopecurus myosuroides*) problems may, however, be best dealt with by soil inversion with the mouldboard plough.

ECONOMICAL CULTIVATION SYSTEMS

A measure of the efficiency of direct drilling and reduced cultivation systems can be gained from the energy input figures for establishing cereals under the various systems. The usual range is as follows:

Direct drilling or zero cultivation	35–80 MJ/ha
Reduced cultivations based on tines and discs	100–230 MJ/ha
Traditional cultivation based on mouldboard plough	200–360 MJ/ha

There are extreme cases which fall well outside these general ranges. For example, many intended reduced cultivations are well into the 'traditional' range and some systems considerably exceed 360 MJ/ha. At the lower end of the range, scratch cultivation for oilseed rape establishment may often be less than 100 MJ/ha, and in the direct drilling range. These figures refer only to the energy that is put into the implements themselves during the course of cereal crop establishment. They intentionally leave out the effort that might be needed to move the tractors themselves over the land. What is obvious from these figures is that direct drilling is easily the most energy-efficient system available to the farmer. The reasons why it has never been of major importance, in spite of this efficiency, are discussed later in this chapter.

In planning an economical cultivation system, an indication of the energy requirement of different implements is of value. Typical figures are as shown in Tables 16.1 and 16.2.

Table 16.1   Energy requirements of different primary implements

Single primary operation*	Typical requirement MJ/ha
Mouldboard plough	166
Heavy disc harrows	93
Chisel plough or heavy cultivator†	93
Direct drilling	53
Shallow rotary cultivation	298

* Primary operation implies that the land has not been previously disturbed.
† Includes heavy duty spring-type cultivators.

Table 16.2   Energy requirements of different secondary implements

Single secondary operations*	Typical requirement MJ/ha
Disc harrows—light	27
Heavy cultivator	53
Spring tine cultivator	27
Corn drill	13
Rotary cultivation†	99

* Secondary operation implies that at least one previous cultivation has been carried out.
† Power required varies widely with rotor design and speed. Half this figure has been recorded for secondary cultivation with the spiked rotor.

In the heavy land situations it is certain that if an economical cultivation system is to be followed, it must be economical at the primary stage of cultivation. If the less efficient implements are used at this stage, there is little scope for making economies at the secondary cultivation stage. On the lightest of soils, however, where primary cultivation energy requirements are much lower than the typical figures listed, it generally works out that the greatest scope for economy is at the secondary cultivation stage.

EFFECT OF CULTIVATION SYSTEM ON TYPE OF TRACTOR

Mouldboard ploughing, at traditional depths and on heavy land, is normally associated with 4-wheel drive and crawler tractors. This is because the draughts are high and forward speeds comparatively low, one factor being that mouldboard plough draught increases considerably with speed (Chapter 9, fig. 9.7). Tined cultivation, even on heavy land, is more often carried out at higher forward speeds with 2-wheel drive and 4-wheel drive tractors. Tined cultivators, discs and combination implements designed for shallow cultivation can operate at higher forward speeds even on heavy land, and 2-wheel drive tractors are generally satisfactory.

Tractor power normally costs most in crawlers and least in 2-wheel drive tractors, so that the change from mouldboard ploughing to tine cultivations and direct drilling has often meant not only less power on the farm but power in a cheaper form. This does not, of course, hold true in every case. There is an enormous amount of mouldboard ploughing carried out with the less expensive 2-wheel drive tractors and, particularly where hills are involved, high power 4-wheel drive tractors may be used in minimal cultivation in order to maintain high working speeds regardless of gradient.

IMPLEMENTS FOR MINIMAL CULTIVATIONS

Heavy Tined Implements

Heavy cultivators or chisel ploughs show very substantial energy savings over the mouldboard plough in their first pass over the land. They are less expensive to buy—about half price on a width basis—demand less skill in use and do not have severe speed limitations. Matching very large tractors with the correct size of implement is easier than with mouldboard ploughs. The main point of importance in the operation of this type of implement is to avoid bringing up large clods which may be difficult or nearly impossible to break without waiting for weathering and which may involve considerable loss of time and effort in moving tractors over the very rough surface produced. This is in fact the area where the comparison of energy input terms on a 'net' basis could be misleading. It is likely that in many cases the effort required to move the tractor over the land is greater than that required to move the implement so that any statement of implement energy requirement may be quite misleading in terms of the final outcome of the day's work.

The mode of operation of the various shapes of tines is set out in Chapter 10. What should be emphasised in terms of minimal cultivation is that although a considerable draught saving is possible by using a shallow approach angle tine, there may be problems in some soils and some seasons of bringing up very large clods. Where this occurs the curved-spring-type chisel plough is probably one of the easiest to control; with care this type of implement can be used to work downwards from the surface in stages under virtually any conditions, so avoiding the problem of producing large unbreakable clods.

Discs

Various forms of discs have a place in economical cultivation systems. Work rate is high, energy input can be low and they have a good mixing action on surface trash. As emphasised in Chapter 10, their action is essentially cutting and compacting so that they can be highly efficient in crushing clods that would otherwise be difficult to trap with a tined implement, or they can do damage (in moist soils) through compaction and localised smearing. The results of experiments using discs in cultivations suggest that they are best used on an opportunity basis rather than as a regular feature of a cultivation system. The incorporation of straw to 100–150 mm in heavy soils calls for weights in the range of 100–150 kg per disc, and tractor engine power of 75–112 kW (100–150 hp) for a 3 m wide implement.

Shallow Ploughing

Energy input into ploughing is closely related to depth of work, and high work rates and low energy inputs have been achieved by reducing ploughing depth. Ploughs designed for work at roughly 200 mm do not generally function well at half that depth, and manufacturers have offered purpose-built shallow or 'skim' ploughs (photo 16.1). These are used in units of up to 10 furrows with furrow width generally about 230 mm and a working depth of about 100 mm. The general experience of farmers has been that provided soil conditions are not too hard, where some of the lighter ploughs may suffer damage to the plough frame, they do a satisfactory soil inversion job at high work rates.

Powered Cultivation Equipment

Power-driven cultivation equipment has the advantage of not

Plate 16.1   Shallow or skim plough

requiring tractor weight for traction, not involving appreciable wheel slip and in being able, where necessary, to do a great deal of cultivation in one pass over the field. Where economy of cultivation is the aim, however, and particularly where cereals are concerned, powered equipment is seldom a good proposition because the power input is high, the work rate is comparatively low and the tilth produced is generally finer than that required for cereals. Some combined power harrows and seed drills have proved valuable in difficult cultivation situations.

SEED DRILLS

Drills are expected to place seed in a variety of tilths, with varying presence of surface straw, to a constant depth, with a highly uniform horizontal spacing, at high forward operating speeds. These requirements are met with differing degrees of success and one single drill is unlikely to meet all conditions successfully.

Where the cultivation treatment does not involve inversion by the mouldboard plough, as much as 30 per cent of the straw crop can remain on the soil surface (photo 16.2). The drill needs to be capable of working through this under wet conditions; if so it will normally cope comfortably with the same amount of straw under dry conditions.

Plate 16.2   Primary discing for straw incorporation   (Parmiter)

Metering and Seed Delivery

The main forms of seed metering mechanism, and the essential elements of seed box, metering units and delivery tubes, have been established for a very long time. Certainly by the middle of the last century the present layout was in common use, while the cupfeed and force-feed systems were known two centuries before that. There is in practice nothing to choose between studded rollers, forms of internal and external feed, and centrifugal distribution systems. With normal cereal sowing rates there is in effect 1/100th of a second between seeds as they leave the metering mechanism and pass down the tube. Therefore a delay of 1/100th of a second means that seeds are being overtaken by others. This delay is related largely to the design of the grain tubes, and can lead to a degree of bunching or certainly a large proportion of doubles in the row of seeds. The grain tubes should be the largest practical bore, not more than 15° out of vertical and telescopic for length adjustment rather than single piece tubes. Given these conditions, 60–80 per cent of the seed will not touch the sides of the tube, and disturbance to the grain flow will be minimal, giving adequate spacing in the row. Curved tubes, which are necessary in wide pneumatic type drills, do cause changes of velocity in the tube, but these are to some extent overcome by the forced air delivery.

Coulters

Coulters may be generally classified as forward-inclined, like the spring tine or hoe type, or backward-inclined like the Suffolk or various forms of disc coulter. The main requirement of the coulter is that it will place the seed on firm, if not undisturbed, soil at the bottom of the opened channel, with minimal bunching of seeds, none landing on the sides of the channel or on dry soil that has fallen into the channel ahead of the seed, and with the channel being finally covered by dry soil. The essential requirements are even spacing, even depth and firm soil contact below the seed to ensure good moisture uptake.

Where there are appreciable quantities of straw on the surface, the forward-inclined spring type machines are likely to block; the same may apply to the very light type of backward-sloping coulter. Under these conditions disc coulters will perform best.

Where soil is loose and fluffy, and this might apply particularly to soils that have had quantities of straw incorporated, furrow pressing may be necessary before drilling, and press wheels on the drill are a distinct advantage. In this way the requirement for a firm base for the seed can be provided.

Overall, there is little difference between drills presently on the market as regards the accuracy of metering and delivery of seed. Most drills will do an excellent job under specific conditions, but some will require freedom from surface trash, fine tilths and level surfaces. Under the majority of commercial conditions today substantial disc drills, often with press wheels incorporated, will be the most successful all-purpose implement.

MINIMAL CULTIVATION AND SOIL STRUCTURE

The most obvious change in soil structure which occurs when topsoils are not loosened regularly by tillage is tighter packing of the soil. This packing takes place at the expense of the larger pores, that is those greater than about 0.1 mm in diameter (Table 16.3).

Table 16.3 Effect of cultivation depth on pore size in soil (per cent by volume in topsoils)

Total pore space		Pores > 0.1 mm		Pores < 0.1 mm	
SC	DC	SC	DC	SC	DC
43	48	10	17	33	31

SC = shallow cultivation (4 cm)
DC = deep cultivation (20 cm)

Larger pores fulfil important functions: water will not drain in their absence, good aeration depends on them, and furthermore root penetration is impeded unless a network of large pores is present.

The changes in bulk soil properties resulting from the withdrawal of tillage therefore appear to be harmful, and the only benefit would seem to be an improved capacity to carry traffic conferred by the greater mechanical strength of the soil. In practice, however, crops commonly refute this conclusion and are unaffected by the conditions! One of the reasons is that water frequently drains faster through undisturbed topsoil due to the greater continuity of such coarse pores as are present. For the same reason aeration is often unrestricted and root growth unimpeded. Larger numbers of burrowing earthworms are invariably associated with undisturbed land enhancing pore continuity. A further reason is that the cracks caused by water extraction and the channels left by crop roots are preserved from year to year.

If soil is left undisturbed or only shallowly cultivated for several years and successful crops are grown, progressive changes take place. The crumb structure or natural tilth of the surface layers improves, largely because of the gradual increase in soil organic matter in the top few centimetres. Earthworm and other faunal activity improves porosity and structural stability in the topsoil and a beneficial organic cycle is developed. But on weakly structured and poorly drained soils conditions may deteriorate progressively as crop growth becomes thin and patchy and weed populations increase. Such divergent trends and the factors responsible for them are illustrated in Table 16.4.

CROP EXPERIMENTS INTO DEPTH OF CULTIVATION

Permanent adoption of new agricultural systems only occurs where the advantages of change clearly outweigh the disadvantages. Reduced tillage is no exception to this. The advantages in saving on labour and time, possibly coupled with savings in machinery investment, have to be balanced with site information on reliability of yield and on weed competition. This balance is further affected by method of trash disposal; now that burning straw is not permitted, the balance tips in favour of deeper tillage.

Cultivation methods in the UK have been comprehensively researched over many years. In 1989 the Home Grown Cereals Authority commissioned a review of this work titled *Reduced*

Table 16.4   Soil conditions in uncultivated land

Factors leading to improvement:

Soil type resistant to overpacking
Good field drainage
Winter cropping
Timely drilling   →   POROUS TOPSOIL
Minimum traffic on field
Tramlines
Dry weather and frosts

Factors leading to deterioration:

Unstable soil type
Poor field drainage
COMPACT IMPERMEABLE   Late drilling
TOPSOIL   ←   Unnecessary traffic and wheel slip
Poor weed control
Excessive surface trash
Prolonged wet weather

Tillage Review No 5, and another HGCA-funded review, *Changing Straw Disposal Practices No 11,* was published in 1990. The findings of these two publications are considered below.

Reduced Tillage in the UK

The aim of field trials into reduced tillage was to examine the potential of faster methods of establishing winter cereals. Most of the experiments compared shallow cultivation (less than 10 cm) and direct drilling with conventional deeper loosening generally by ploughing. The duration of individual experiments ranged from 1–20 years. Limitations other than soil conditions were as far as possible avoided. Thus straw was usually burnt and weeds, diseases and pests were controlled by overall treatment coupled with individual plot treatment where necessary. Cultivation treatments at a site were all drilled on the same day wherever possible. The mean yields from all the trials are shown in Tables 16.5 and 16.6.

The results demonstrate that carefully managed minimal cultivations give very similar cereal yields to ploughing. In commercial practice minimal cultivation is rarely advantageous on sands and light loams, and ploughing is the best option except perhaps on very large farms where speed may be the first priority. It is the claylands which benefit most from non-plough tillage, partly because of the large draught of ploughing but also because of the better seedbeds which result from not burying the natural surface tilth (photo 16.2). This factor is particularly relevant to the establishment of small-seeded winter crops such as oilseed rape.

Table 16.5 Mean yield of direct drilled winter and spring cereals relative to ploughing. Average yield after ploughing = 100

	1967–75	1975†	1969–74
Spring barley	93 (61)*	87 (17)	95 (44)
Winter cereals	96.3 (57)	95.8 (28)	97 (29)

* Number of trial years in brackets
† Abnormally wet autumn-spring period

Table 16.6 Mean yield of minimal cultivated cereals relative to ploughing. Average yield after ploughing = 100 (1969–86)

	Shallow cultivated	Direct drilled
Clays	101.5 (81)*	101.3 (85)
Medium loams	100.8 (88)	99.2 (88)
Light loams and sands	101.3 (42)	98.8 (42)

* Number of experiment years in brackets (i.e. number of sites × number of years)

Cultivations for Straw Incorporation
In most of the experiments covered in the previous section, trash problems were avoided by burning the straw. But as from 1992 this option is no longer available in the UK. There is no doubt that minimal tillage is less attractive where straw cannot be burnt, and the case for deeper tillage with plough, discs or cultivator is increased.

A national series of field experiments to examine cultivations for straw incorporation was started by MAFF and the AFRC in 1982. Most of the work was with winter cereals. Results in Table 16.7 show that ploughing came out top for all soils although for heavy clays the disadvantage of shallow cultivation was very small.

Choice of cultivation and method of straw disposal should not be considered in isolation from other aspects of crop husbandry. Weed competition and weed control are strongly influenced both by type of tillage and by method of straw disposal. The strongly competitive grass weeds and volunteer cereals are favoured by both non-inversion and straw incorporation. Table 16.8 shows the percentage herbicide control required to stabilise a blackgrass population in various husbandry situations.

Table 16.7   Mean yields of winter cereals following incorporation of chopped straw by cultivation compared with ploughing

	Straw incorporation by disc/tine* to 10–15 cm depth
Heavy clays	99.5
Medium clays	98.5
Medium loams	98.0
Light loams and sands	90.5

* Avoiding experiments in which grass weeds could not be satisfactorily controlled.

Table 16.8   Percentage kill by herbicides needed to maintain a static population of blackgrass

	Straw burnt	Straw incorporated
Ploughed	50	65
Shallow cultivation	88	92

Source: G. W. Cussans, Rothamsted

These figures emphasise the increased weed pressures likely to be encountered by sequential shallow cultivation particularly when straw is incorporated. However where cereals are interspersed with break crops this pressure is less important. Similar effects are noted with volunteer cereals, but they can reduce the acceptability of milling and seed wheats and of malting barley.

Cereal diseases and pests are much less affected by choice of cultivation and method of straw disposal than weeds. Normally fungicide use need not be modified according to cultivation type or method of straw disposal, and although incorporation of straw probably encourages slugs, quality and consolidation of seedbed seem to be more important factors in determining the incidence of damage.

Nitrogen requirements of crops is so little modified by both cultivation type and method of straw disposal that ADAS do not vary nitrogen recommendations for these factors except in the case of oilseed rape grown on clays. Where straw is incorporated before this crop, seedbed nitrogen is advised but not when straw is baled.

Soil Type and Minimal Cultivation

Certain characteristics of soil encourage the open porous conditions which favour minimal cultivation. Natural free drainage and natural restructuring are two of the properties involved. A high natural lime content is also beneficial.

Cotswold Brash, magnesian limestone and shallow chalk soils are all good examples of soils suitable for shallow tillage. However where there is need to incorporate straw, ploughing is likely to be the best option in most situations. All the calcareous clays have a stable structure and readily restructure when compacted. Soils of this type, for example Hanslope series on chalky boulder clay and Wicken series on Lias clay, respond well to shallow cultivation with deeper loosening every few years. Even where straw has to be incorporated, ploughing is not the best option for clays except when grass weed competition is high, compaction is severe or residual herbicides may damage the next crop.

At the other extreme weakly structured sandy soils and light silt soils easily lose structure and normally respond well to annual loosening of the topsoil.

GUIDELINES FOR MINIMAL TILLAGE

Unnecessary or recreational tillage is widespread on farms even though every pass costs money. The difficulty of clearly defining when cultivation is necessary is one of the reasons for this unproductive work. In part this is because the interaction between crop growth and soil conditions is complex and dependent on weather, and in part because cultivation has indirect effects on weed competition.

Establishment of Winter Oilseed Rape

Where large areas of rape are grown, shallow tillage with discs or heavy duty spring tines is often the best option if compaction is absent or slight. On clays, consistent timely crop establishment is not possible after ploughing, and very shallow loosening after baling is invariably the best option, not only because of the faster work rates but also because of the finer and moister seedbeds which can be achieved. However in the presence of chopped straw, shallow cultivation has given less consistent results. Incorporation of the straw calls for cultivation to at least 10 cm with inevitably greater loss of seedbed moisture. Rolling is always a key element of shallow tillage and is even more vital where straw is incorporated. Well chopped, evenly spread straw and a short stubble (15 cm or less) are important ingredients of success.

Alternatively a broadcasting technique which has proved satisfactory at ADAS Boxworth should be considered. After the straw has been chopped and spread, moisture tends to build up under the mulched straw. This is exploited by broadcasting the rape seed and cultivating about 2 cm into the soil with a rotary harrow or equivalent implement prior to rolling. This technique has given good results but tends to be less consistent in dry seasons.

A third possibility of drilling direct through the chopped straw with purpose built drills designed to provide sufficient coulter pressure and separation of straw from drilled seed needs further development.

Establishing Winter Cereals on Clays

The results of the national trials summarised in Table 16.7 demonstrate that clayland farmers should not automatically assume that straw incorporation always requires the plough. Although incorporation increases the optimum working depth of cultivation, good crops can be achieved without resort to ploughing every year. However if the previous crop is extensively lodged, shallow incorporation should not be attempted. Stubble should be cut as short as practicable and no longer than 15 cm. The finer the straw chop and the more uniform the spread of straw and chaff, the easier incorporation becomes. Incorporation should not be delayed beyond the time needed for straw to settle through the stubble onto the ground. The first shallow incorporation with discs or heavy duty flexible tines leaves a great deal of straw on the surface but starts the process of reduction in bulk of straw. This process is hastened by rolling. After 10–14 days, disc or tine cultivation should be repeated at an angle to the first work and to about 10 cm depth, followed again by rolling. If the chop and spread has been good this amount of incorporation is often sufficient. The criterion to use is suitability for drilling: if the straw does not drag up on coulters and the seed is drilled deep enough and evenly, the presence of substantial amounts of straw is of no consequence. Disc coulter drills will cope with more straw in the surface than Suffolk coulters.

For those who do not want to plough and where shallow

incorporation is not a satisfactory option, deeper cultivation is a useful alternative. The success of this system is based around either heavy discs or rigid tines with twisted shanks. Again, rolling after each pass to preserve moisture is important in dry autumns. Following two disc cultivations 10–14 days apart, a heavy duty rigid tine cultivator should be used to loosen just below the depth of discing.

CLIMATIC SUITABILITY

In most cases it is wet rather than dry conditions which are responsible for problems in minimal tillage. Drainage problems are shown up by wet periods and all soils are more susceptible to compaction in wet conditions. Therefore it follows that minimal tillage is a technique most likely to be successful in drier climates. This is certainly the case with sequential shallow tillage, and for winter cropping winter cereals, a wetness index—average return to field capacity (RFC)—has been suggested for delineation of areas favourable for this practice. The division has been put at an RFC date of 1st November which includes the major cereal-growing areas of eastern, central and southern England. For information on RFC dates see MAFF Bulletin 34. In wetter areas naturally free soil drainage is necessary for successful minimal cultivation.

CHAPTER 17

EROSION

Erosion of land by wind and water has destroyed thousands of hectares of good agricultural land in other countries. Probably the best known of such disasters is the 'dust bowl' which formed in parts of the central USA in the 1920s. It developed following the widespread practice of continuous arable cropping of land which had previously been grassland or forest.

Continuous arable cropping has become established in many parts of Britain particularly in the east, and in spite of predictions that disaster would follow there is as yet no English dust bowl. But could it happen here? Some pundits point to the occurrence of occasional wind erosion on peat soils in the Fens and on some sand lands, claiming they are the beginning of such troubles. The purpose of this chapter is to discuss the principles governing erosion, the measures which can be used for its control, and the risk of erosion in this country.

PROCESSES OF WIND EROSION

Wind erosion, often known as 'blowing', starts with a process known as 'saltation', which is a sort of jumping motion of the smallest, most erodible particles or aggregates of soil. Particles which become airborne in gusts of wind fall gradually back to the ground in a trajectory rather like that of a bullet after covering a distance of three or four metres. The impact of these particles hitting the ground moves other particles of a similar size and also larger particles. The larger particles slide and roll along the surface in a process known as 'avalanching' until stopped by ditches or hedges (see fig. 17.1 and photo 17.1) while finer material is

Plate 17.1   Wind eroded sand collecting in a Suffolk ditch
(Arable Farmer)

projected higher into the air to form dust clouds. The erosion
process sorts the soil particles according to their size, moving the
coarser ones only short distances, while depositing the finest much
greater distances, often as much as kilometres downwind. With
successive blows the eroded soil becomes more coarse as most of
the fine material is blown away.

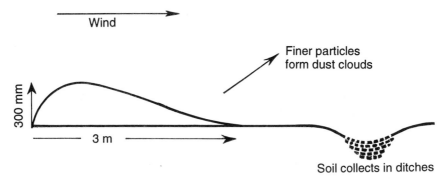

Figure 17.1   Effects of wind erosion

Factors Affecting Wind Erosion

The velocity of the wind at the soil surface is one of the most important factors affecting erosion. A speed of about 30–40 km/hour is needed before blowing starts. Moist soils are much less liable to erode than dry ones because moist particles stick to one another to some extent. A warm dry wind which rapidly dries out the surface is much more liable to give trouble than a cool humid one, and soils in low-rainfall areas are more susceptible to blowing than soils in wet areas.

The topography is important since it affects wind speed. Flat or gently undulating land is more susceptible to erosion than hilly country which slows down the speed of the wind.

Bare land is much more susceptible than cropped land because the stems and leaves of plants slow down the surface wind speed, and the roots of closely drilled crops such as cereals tend to bind the soil together. The roots of widely spaced row crops such as sugar beet give the least protection. A loose surface with a fine tilth gives the greatest opportunity for the wind to start erosion, and a coarse cloddy surface is least susceptible.

In mineral soils the most erodible particles are the fine sand fraction with diameters in the region of 0.1–0.2 mm. On peat soils, which have a much lower density, the most erodible particles are about 1 mm in diameter; these are small aggregates and can be formed by natural weathering.

The extent to which a soil is aggregated, the size of the aggregates and the ease with which they can be broken down by natural weathering, cultivations and abrasion from eroding particles, all have an influence on the erosion risk of a soil. Light mineral soils in arable cropping generally have very weakly developed structures and their erosiveness depends on the amount of fine sand they contain. Silt, clay and organic matter all tend to increase aggregation in soils and so reduce susceptibility to erosion. In peaty and organic soils, aggregates have such a low density and structure is so loose that the high organic content (greater than 25 per cent) increases the risk of wind erosion.

Damaging wind erosion in the UK is confined to exposed sandlands in the Vale of York, Nottinghamshire and parts of East Anglia including the peaty Fenland soils.

CONTROL OF WIND EROSION

Wind erosion can be completely prevented by having a permanent cover of grass or trees which protects the soil from the wind. If the land is cropped then it will be susceptible to erosion when the land is bare during the spring and winter. Erosion control should aim to:

1. Reduce the wind velocity at the soil surface.
2. Stabilise the soil surface.
3. Trap soil particles already moving.

Shelter Belts

Shelter belts or windbreaks of trees or hedgerows reduce wind velocity in their lee and also trap airborne soil. For maximum protection to leeward, the shelter should not be impenetrable but should allow about a third of the wind to pass through and interfere with the wind curving downwards over the top of the shelter. Impenetrable shelter belts can cause turbulence immediately to leeward, which may make erosion worse. The alignment should be at right angles to the direction of the prevailing winds, thus accepting the risk of occasional blows from other directions. Shelter belts give protection for a distance of about twenty times their height. Thus a 5 m high belt will give protection for only up to about 100 m. Only very high-value crops can support the cost of maintaining shelter belts every 100 m.

Strip Cropping

The strip cropping system, in which strips of row crops are alternated with strips of grass or cereals which are resistant to erosion, is widely used in the USA. The grass or cereal strip traps moving soil, reduces the surface wind speed and keeps avalanching in check, provided the width of erosion-susceptible crop is sufficiently narrow. On highly erodible soils strips of about 50 m width are necessary which creates great difficulties in farm management. The method also suffers from the same directional problem as shelter belts and has been little used in this country.

Cover Cropping

Establishment of a fast growing cover crop in advance of the cash crop, is a system successfully used on peats and sands in eastern England. The protective crop is usually spring barley, rye or mustard (colour photo 9), drilled several weeks before the crop requiring protection. It affords protection against 'sand blasting' by eroding particles, and also protects the soil by slowing down the surface wind speed. The system works well for wide-row crops in which the nurse crop can be selectively killed by a herbicide in the growing cash crop. However, the nurse crop can reduce yield by competition and its destruction thus needs to be timed with care.

Straw Planting
A technique of 'planting' long straw between the rows of susceptible crops before drilling successfully avoids the problems associated with living shelter crops. The method requires a specialist machine and is comparatively expensive. It is attractive for very early sown crops, such as onions, where living cover cannot be established soon enough.

Mulches
Manures, sewage sludge and sugar-beet factory lime sludge can be used to protect the soil surface. However, possible adverse effects from frequent use need to be considered, for example the build-up of toxic metals from sewage sludge or the risk of over liming from sugar-beet factory lime sludge.

Synthetic Stabilisers
Synthetic anticapping materials such as polyvinyl alcohol can effectively protect the surface of sandy soils. Their use is economic when high-value crops are grown. They are not successful on peat soils because cracking and movement of the soil makes them ineffective.

Claying and Marling
One of the most effective ways of preventing blowing is the practice of claying or marling. However, large quantities need to be applied—at least 400 tonne/ha—so the method is only economic where the clay or marl is available on the same farm or can be obtained free from, for example, a major drainage scheme. On peat soils which overlie clay, subsoil mixing is one way of bringing clay to the surface.

Cultivations
The careful selection of type, frequency and timing of cultivations can be very effective in controlling erosion of both sands and peats. On sandy soils the erosion risk increases with the number of cultivations and consequent breakdown in structure, so the aim should be to keep cultivations to the absolute minimum. When the soil needs to be loosened, tined implements rather than the plough should be used wherever possible, so give the soil some protection against erosion by keeping crop residues on the surface. Another

technique involves ploughing late when the soil is moist to give a cloddy tilth, followed by seedbed cultivations confined to the drill rows.

Another method successfully developed on the sands at ADAS Gleadthorpe is to plough in spring with a furrow press to provide surface consolidation. Crops are drilled directly into the pressed land, usually at right angles to the direction of pressing, and the surface remains resistant to erosion (colour photo 10). This technique is not effective on the lightest East Anglian sands which contain insufficient fine particles to hold together at the surface. A disadvantage of this method arises where slopes are steep enough to cause water erosion. In such situations the furrow press marks tend to aggravate run-off.

On peat soils crop residues on the surface reduce the risk of erosion, but natural weathering caused by wetting and drying lifts the surface and lets in the wind. Freshly ploughed and cultivated land, left rough with damp clods brought to the surface, is much less susceptible to blowing than a smooth surface. A common practice takes advantage of this by pulling tines or hoes through bare land when strong winds are forecast.

PROCESSES OF WATER EROSION

In much the same way that wind erosion is caused by rapidly moving air, water erosion is caused by water moving rapidly across bare soil. As it flows downslope, water takes into suspension first the clay and silt and as it picks up speed, so it takes up progressively coarser material. Three more or less distinct forms of water erosion are recognised: sheet, rill and gulley. Sheet erosion removes soil fairly uniformly all over the slope but usually only in small amounts. In rill erosion numerous shallow gullies or 'rills' are formed. When water is concentrated into streams deep gullies are formed and large losses of soil occur.

The beating action of rain falling on wet soil destroys surface aggregates and so reduces the rate of infiltration of water. Once the infiltration rate is exceeded by rainfall, surface run-off takes place down the slope. The longer and steeper the slope, the greater becomes the erosive power of the run-off. Aggregates in clays are comparatively strong and much less prone to breakdown; low organic matter sandy and silty topsoils have weaker aggregates and are consequently more susceptible to erosion. Anything which decreases the rate of percolation, e.g. compaction (photo 17.2), or speeds up surface water movement, e.g. a smooth surface, increases the risk of erosion. Grass or narrow row crops like cereals protect the soil by root binding of the surface layer, while wide-row crops such as sugar beet are slow to provide surface protection and are more susceptible to erosion.

EROSION IN BRITAIN

Prior to the 1960s water erosion in Britain was generally thought to be rare and confined to a few well-known sites. However it is now clear that soil loss from this cause has become more common and widespread in recent decades. There are probably several reasons for this: traditional grassland has given way to arable farming in many areas, field size has increased at the expense of strategically sited hedgerows, and uncropped tramlines, which channel run-off, are now used throughout arable areas. The large increase in winter cereal growing during the 1970s also led to more erosion. The reasons are fine seedbeds and frequently low crop cover overwinter coupled with uncropped tramlines orientated up and downslope.

Severe erosion, as seen in countries often experiencing high intensity rainfall, is uncommon in Britain and is confined to cloud-burst type storms (colour photo 8). However, slight to moderate soil

Plate 17.2   Water erosion over a plough pan on a sandy field in Bedfordshire.

loss is common in areas with susceptible soils (photo 17.2).

Measurements of soil loss by erosion were made in 18 extensive traverses of susceptible areas of lowland England from 1981–84, using aerial photography complemented by ground visits. Most of the erosion identified in this study took place on free-draining, easily worked soils, for example West Midlands sands and the chalk soils of the South Downs.

Based on the evidence of various studies of erosion including the one mentioned in the last paragraph, Dr Evans concludes in an article in *Soil Use and Management*, Volume 6 1990, that three-quarters of the land surface of England and Wales has negligible risk of erosion, nearly a fifth is at moderate risk and about one-twentieth at high risk. High risk is defined as more than 5 per cent of fields affected every year. Studies in arable parts of Scotland also suggest that water erosion is now more common than earlier in the century and probably affects a larger proportion of agricultural land than in England.

Although it would be misleading to suggest that water erosion is a major hazard to British agriculture, there is no doubt that locally it has become an important problem with consequences both for the gradual decline in soil fertility and also for the pollution of the non-agricultural environment. Off farm problems arise from deposition of eroded soil on roads, in watercourses and occasionally inside houses. The escape of soil into streams and rivers results in organic nitrogen phosphate and pesticide contamination, but the extent of this problem in Britain and its significance relative to other sources of these contaminants have not been comprehensively examined.

CONTROL OF WATER EROSION

Husbandry to reduce soil loss by water erosion aims to minimise run-off and to reduce the damage caused by any run-off that is generated. In situations where severe gulleying occurs regularly in the same field, the only secure control measure is to establish permanent grass or forestry. However, only a small proportion of susceptible land requires such drastic treatment and in the majority of cases comparatively small changes in management can greatly reduce the risk of losing soil.

Erosion from Land in Winter Cereals
Although well established cereals provide protection to land,

much of the erosion in the UK takes place from winter cereal land. The typical situation is where above average rain falls on tight fine structured seedbeds with negligible crop cover. Faster cultivation allowing earlier crop establishment is the most important change required. Chopped straw worked into the surface provides useful protection during the period of establishment and care must be taken not to produce seedbeds that are too fine and tight. The practice of rolling with heavy rolls to bury stones is best delayed until spring and only lighter rolls used to ensure consolidation during dry periods. The problem with run-off generated by tramlines can be reduced by drilling overall rather than leaving rows undrilled; tramlines are marked at the first time of spraying when the crop is already providing some protection to the surface.

Erosion from Land in Sugar Beet and Potatoes

On sands at ADAS Gleadthorpe in Nottinghamshire, winter rye is established in September on ploughed land and provides protection against water erosion and nitrate leaching over winter. The rye is killed off with a herbicide immediately prior to drilling the beet with a drill designed to penetrate trash. Although this technique needs careful management it is proving to give much reduced soil loss when tested with rainfall simulation.

Good erosion control is difficult to achieve in potato fields particularly where irrigation is practised. A system of 'tied' ridges which block off furrows at intervals downslope, together with the 'Dammerdyke' which scoops out small holes along wheelways to provide temporary water storage, will provide partial control.

Other Control Measures

Damaging run-off is sometimes generated by tracks and farm roads or even by ponds. Interception ditches are needed to prevent surface water from these areas reaching fields. Hedgerow removal has created many fields with long slopes generating run-off. Re-establishing hedgerows or permanent grass strips across slopes can substantially reduce erosion losses.

Periods of severe erosion on farms are often interspersed with years with few problems which tend to lull farmers into a false sense of security. Consequently it is essential to practise control measures as a routine every year even though they will often be unnecessary.

Chapter 18

POLLUTION AND THE SOIL

The physical effects of modern farming on the soil and soil structure have been thoroughly discussed in the previous chapters. However, the implications of today's bewildering array of pesticides on the biological processes in the soil are just as worthy of consideration. Is there a danger of destroying the many forms of life in the soil and making it sterile? What happens to chemicals in the soil? Are fertilisers polluting our rivers? These and other questions are discussed in this chapter.

HERBICIDES

It has been found that soils possess remarkable powers of breaking down synthetic organic pesticides. Soil micro-organisms such as bacteria and fungi need carbon for energy in much the same way as animals need carbohydrate. Normally they get carbon from plant and animal remains, but many soil micro-organisms are able to adapt and use the carbon in herbicides. Often there is a time-lag while the organisms adapt themselves to the new source of food. Some herbicides are broken down in a matter of days, while others are more persistent, particularly soil-acting herbicides, some of which may take a year or more to disappear completely. Thus there may be a hazard to the crop following the one to which a soil-acting herbicide was applied. However, all the evidence indicates that there is no long-term hazard either to crops or soil micro-organisms. Many times the recommended dose must be applied to damage the soil's population of bacteria, fungi and other micro-organisms.

Traces of herbicide above the EC limit of 0.1 microgram per litre are being detected in UK water supplies. The limit is so strict that the excess is of no known danger to health but it does break EC law. By far the most commonly detected herbicides are atraozine and simazine used for industrial weed control but some herbicides used only on crops, particularly products applied from autumn to early spring to cereals, are also being found. Apart from avoiding spillage into watercourses there is little that farmers can do to avoid this pollution except for using low dose products when suitable ones are available.

INSECTICIDES

The insecticides that have caused most concern as pollutants are the synthetic organic compounds which came into use during and since World War II. The long persistence of some organo-chlorine insecticides, particularly DDT and dieldrin, has been well publicised. They were very effective and useful insecticides but their widespread use resulted in accumulation in the soil, food, wildlife and in the global environment. It also resulted in certain pests becoming resistant to these materials, for example, carrot fly on the fen peat soils, and in widespread kill of predators, thus upsetting the pest-to-predator balance. For these reasons, and because of the uncertainty of the long-term effects, their use in Britain is now completely banned. They have been replaced by the less persistent organophosphorus, carbamate and synthetic pyrethroid insecticides, which are generally metabolised rapidly by animals and excreted.

The organophosphates are broken down completely by weathering or the action of micro-organisms within weeks or a few months, and so tend not to be very persistent in the living environment. There is no adverse effect on soil micro-organisms, but their effect on the soil fauna is less clear. It is known that they can kill earthworms and some small soil insects, but there is no evidence that this affects soil fertility. Obviously, though, it is important to avoid excessive or careless use of such chemicals which because of their acute toxicity need to be handled with caution.

Insecticides and nematicides applied to the soil to control soil insects and nematodes are potentially the most damaging to soil organisms. Some of the nematicides are particularly poisonous chemicals. However, they are short-lived and as they come into contact with only a small part of the soil volume they do not accumulate and have no significant or lasting effect on soil biology.

Insecticides have been detected in water supplies very infrequently because most are applied during the summer when little leaching occurs and the dose for many of them is extremely small.

FUNGICIDES

Although toxic to fungal spores, fungicides have a very low level of toxicity to animal life and present no hazard to the life of the soil. Most have only a short persistence. Residues have rarely been detected in water supplies.

HEAVY METAL TOXICITY

The effects of soil contamination with heavy metals which are toxic to plants in large amounts can be much more damaging and permanent than those of organic pesticides. Metals are strongly absorbed by the soil clay and humus and so do not leach to any extent. Once a soil is contaminated by metals, it remains so indefinitely.

Metals can reach the soil after emission from the chimneys of metal refineries, but the area of land likely to be affected in this way is not very extensive. Metal pollution of agricultural land in Britain occurs mainly through the use of contaminated sewage sludges from industrial towns. Another potential source of contamination with copper and zinc is the application to the land of slurry from pigs fed on a diet containing high concentrations of these metals.

Metals in sewage become adsorbed by the solid matter and so are retained in the sludge. The liquid effluent which is eventually discharged to the rivers has a very low metal content. So the solid sludge, which is of potential value to the farmer as an organic manure rich in nitrogen and phosphate, is also a potential hazard when metals are present in large amounts.

Zinc
Zinc is by far the commonest metal contaminant in sewage sludges. It is a minor plant nutrient and is present in all vegetable foods, but it is not retained by the body, so that all sludges 'naturally' contain a significant amount of zinc. Paper and talcum

powders contain zinc, so sludges even from non-industrial towns tend to be somewhat contaminated with this metal. The levels, though, are usually not high enough to render them unsafe for use on agricultural land at the recommended rates of application.

The main sources of contamination with zinc are metal-plating factories which discharge zinc containing waste into the sewers. Sludges from highly industrial areas can often contain extremely high levels in which case they are quite unsafe for application to agricultural land.

Copper
Copper is the second most common metallic contaminant in sewage sludges, and the main sources of this metal are also metal-plating factories. Plants require and contain much less copper than zinc and consequently sludges from non-industrial or rural areas have low copper contents. However, it is rather more toxic to plants than zinc.

Nickel
Nickel is a serious contaminant in sewage sludges from some towns with nickel-plating works or factories producing products containing nickel, such as needles. It is several times as toxic as zinc to plants; consequently it is very important to avoid application to the land of sewage sludges containing high levels of nickel.

Chromium
Chromium can be present in sludges from towns containing chrome-plating works or tanning industries, but it is much less toxic to plants than the other metals, and is therefore only damaging at very high levels. However, there is some evidence that it may have adverse effects on human health, so it is an undesirable contaminant.

Crop Sensitivity
Crops vary considerably in their sensitivity to metal toxicity. Most horticultural crops and root crops are more sensitive than cereals which, in turn, are more sensitive than the grasses. Although less sensitive than arable crops, the better grass species tend to be replaced by weed grasses such as *Poa* and *Agrostis* on metal-contaminated land, with the result that the level of production is reduced.

Minimising Toxicity

Once the soil is contaminated there is no known way of removing metals from it other than by physically removing the whole topsoil. The toxicity of zinc, nickel, and to a lesser extent copper, is less when the soil is limed than when it is acid. So the adverse effects can be minimised by generous liming, but they cannot be overcome completely if the level of contamination is high. As contamination is irreversible, it is vitally important to avoid applying excessive amounts of metal to the soil. Only sludges known to have a safe metal content should be used, and when these are applied the metal content of the soil must not be allowed to exceed the safe limit as explained below.

Other Toxic Elements

Besides the more common contaminants–zinc, copper, nickel and chromium–sludges from some industrial towns may also contain smaller amounts of cadmium, lead and mercury, all of which are very poisonous to animals and human beings. Cadmium is also toxic to plants and is readily taken up into the leaves and other parts. Consequently it is a particularly dangerous contaminant. There is also the risk of grazing animals being affected by metal poisoning through eating fodder contaminated with soil. The amount of mercury applied in a contaminated sludge could be many times greater than the extremely small quantities which used to be applied in seed treatments.

It is necessary to limit strictly the total amounts of such metals that are applied to land in sludges. The maximum amounts of several common toxic elements that can safely be applied to uncontaminated land are 5 kg/ha cadmium, 500 kg/ha lead, 2 kg/ha mercury, 4 kg/ha molybdenum, 5 kg/ha selenium and 10 kg/ha arsenic. If several applications of sewage sludge have already been applied to a field it is advisable to have the soil analysed before further amounts are applied.

Safety Limits for Heavy Metals

Maximum allowable amounts of toxic metal which can be applied per year and maximum allowable concentrations in sludge are stipulated in a 1989 Statutory Instrument (Table 18.1).

The Statutory Instrument also stipulates limits to the amount of heavy metal allowed in soils (Table 18.2).

Table 18.1 Maximum amounts of toxic metals which can be applied and maximum concentrations in sludge

Heavy metal	Kilograms per hectare per year	Parts per million in sludge dry matter
Zinc	15	50
Copper	7.5	25
Nickel	3	10
Cadmium	0.15	1
Lead	15	25
Mercury	0.1	0.1

Table 18.2 Maximum amounts of heavy metals allowed in soils

Heavy metal	Limit (parts per million in dry soil) according to pH of soil			
	5.0–5.5	5.5–6.0	6.0–7.0	above 7.0
Zinc	200	250	300	450
Copper	80	100	135	200
Nickel	50	60	75	110

	For pH 5.0 and above
Lead	300
Cadmium	3
Mercury	1

POLLUTION OF WATER BY NUTRIENT LOSSES FROM THE SOIL

Water draining through a soil under grassland, arable or forest has always contained small amounts of plant nutrients. Nitrate is normally present, together with small amounts of potassium, magnesium and sulphate, but phosphorus is virtually absent because phosphates are rendered insoluble and so are retained by the soil. Water draining from arable land usually contains more nitrate than that from grassland or forest irrespective of whether nitrogen fertiliser has been applied. Nitrate formed from breakdown of soil organic matter, together with any fertiliser not absorbed by crops, is leached out of the soil during the autumn and winter months except in the drier parts of the country where leaching is often incomplete.

The amounts of potassium, magnesium, and sulphate leached from soils are too small to be of any consequence, but the amount of nitrate can be if it results in a high level of nitrate in rivers and lakes. The nitrate content of water in lakes and reservoirs is a factor in the process of 'eutrophication' which is described below. The nitrate content of water used for drinking is important because it effects the health of babies. If the level rises above about 20 parts per million of nitrate nitrogen, there is a risk of babies suffering from a disorder known as methaemo-globinaemia. This can be fatal in severe cases caused by highly contaminated water. Only babies are susceptible to this complaint and adults are unaffected by quite high levels unless part of the stomach has been removed surgically.

The nitrate levels in British rivers have increased in recent years, and in some English rivers average concentrations are well above the World Health Organisation recommended limit of 22.4 parts per million of nitrate nitrogen. However, there has been little or no further increase since the late seventies. Levels vary greatly with time of year, tending to be highest in winter and early spring when water is draining through the soil, and lowest in the summer when it is being used up by plant life in the water.

It appears that nitrate leached from agricultural land provides a major part of the nitrate in rivers. Since, though, the rates of nitrogen applied by farmers to arable crops are now about the optimum, this source of nitrate in rivers is unlikely to increase. There is still scope for increase in nitrogen usage on much of our grassland. Although grass is a very efficient absorber of nitrate there is considerable nitrate loss from grazed pastures treated with high rates of nitrogen, so increased nitrogen usage would further raise nitrate levels.

A survey of over 600 private boreholes in Britain revealed that the nitrate nitrogen level of the water from four of these exceeded 20 ppm, while all except twenty-eight had levels of less than 10 ppm. Levels tended to be higher in the drier eastern counties than in the west. Although only a small percentage of the total had high levels, nevertheless there is some cause for concern.

Farmers should help to minimise nitrate pollution by sound practices. They should avoid excessive rates of nitrogen fertiliser which, in any case, are a waste of money. In wet springs particularly on sandy soils nitrogen can be leached from the soil and wasted if, with the object of spreading the work load, it is applied too soon before it is needed by the crop. The practice of applying all the nitrogen to the seedbed for spring cereals sown in February on sands can result in its complete loss in wet springs. Such early sown crops should on these soils be regarded as winter crops, and the bulk of the nitrogen should be applied as a top dressing, when growth commences in the spring.

Several areas of land overlying important aquifers where water is extracted from boreholes in, for example, sandstone or chalk have been designated as Nitrate Sensitive Areas. Farmers in these areas are asked to limit nitrogen usage and are compensated if they contract to adhere to the following guidelines:

- Apply no excess fertiliser nitrogen.
- Apply no fertiliser nitrogen in autumn and winter.
- Apply no extra nitrogen for breadmaking wheat.
- Avoid leaving ground bare overwinter by growing cover crops.
- Limit the annual use of animal manures to 175 kg/ha total nitrogen.
- Apply no poultry manure or slurry in late summer and autumn.
- Plough up no long term grass.

If these guidelines are followed it is believed that water draining into aquifers will not contain excessive amounts of nitrate.

EUTROPHICATION

Water in upland streams and reservoirs filled by them contains very low levels of nitrate and phosphate. If the content of these nutrients is raised, an extensive growth of algae can develop in lakes and reservoirs during the spring and summer. This formation of floating green algae is known as 'algal bloom'. The algae, which are tiny plants, sink to the bottom when they die and oxygen is used up as they decompose. The 'blue-green' algae are common in algal blooms, and when these decompose they emit toxins which can kill fish and animals.

This process of enrichment of waters by nutrient salts is known as 'eutrophication', and the effects which follow can cause a complete kill of fish.

Recent evidence suggests that streams flowing through arable land have always contained sufficient nitrate nitrogen to support algal growth, but that the phosphate levels were insufficient because this nutrient is not readily leached out of soil. However, the increasing proportions of sewage effluent, which is rich in phosphate derived mainly from household detergents, has resulted in appreciable increases in the phosphate contents of some of our rivers. This is obviously an important contributory factor to the problem of eutrophication where river waters are abstracted for storage in reservoirs.

CHAPTER 19

THE CLASSIFICATION OF AGRICULTURAL LAND

Soil survey maps show areas of similar soils, which are called 'soil series'. Areas of soils with similar horizons or layers developed from similar materials are included in the same series. As an example, the Newport series, which is found in the Midlands, the North and East Anglia, consists of a loamy sand or sandy loam topsoil over a reddish-brown sandy subsoil below which is sand. The colour of the subsoil is fairly uniform without mottling, indicating that it is a freely drained soil. All soils which fit this description developed on sand are included in the Newport series.

Another example of a soil series is the Windsor series which is found in and around the Thames Valley developed on London Clay. It consists of clay loam or silty clay loam overlying clay. The clay is mottled brown and grey indicating that the drainage is poor.

Soil series are given names as an aid to recognition, the name normally being that of a village or town near to where the soil was first recognised. However widespread a soil series may turn out to be, it keeps the same name. It should be emphasised that, although a soil series comprises soil profiles which are similar, they are nevertheless not identical, and every soil series includes a range of properties for each horizon within defined limits. The amount of detail that can be included on a soil map depends on the scale of the map and also on the number of soil examinations which the surveyor makes which obviously cannot be unlimited. Most soil maps published in this country are 16 mm or 40 mm per km, and areas mapped as a soil series will inevitably contain a small proportion of soils of another series. Very variable areas, in which the soil series changes over short distances which are too small to

261

be separated on a map, are recorded as 'complexes' of two or more series.

Some of the earliest soil surveys were in fruit-growing areas, for example in west Cambridgeshire, the Wisbech area, the Vale of Evesham and south Hampshire. These surveys successfully defined the soil requirements of the various fruit crops. The surveys showed that since they have permanent root systems, fruit crops will not tolerate poor drainage. However, some crops, for example plums and pears, were shown to be able to stand slight drainage impedance, while raspberries and cherries must have well-drained soils.

Recent surveys (figs. 19.1 and 19.2) describe all readily observed features of the soil and can be useful for purposes besides agriculture, such as forestry and land use planning. Soil Survey maps produced by the Soil Survey and Land Research Centre, Silsoe Campus, Silsoe, Bedfordshire, provide invaluable information to agricultural scientists. In addition to these detailed maps of relatively small areas a set of six national soil maps at a scale of 1:250,000 are available. Since most users are interested in crop suitability some interpretation of the map is desirable.

LAND CAPABILITY

The range of crops which can be grown in a field does not depend only on the soil type. The climate has a marked influence; e.g. clay soils which are suitable for mainly arable cropping in eastern England may be restricted almost wholly to grass in the wetter west. Sandy soils of limited value due to drought in the east may be very productive in a high rainfall area. Although a first-rate farmer may be very successful on difficult land, it is not possible to completely overcome limitations. Attempts to introduce intensive cropping with roots on to land only suitable for cereals, or to grow continuous cereals on land suited only to ley arable farming, are bound to fail eventually.

Land Capability Classification
As an aid to fitting suitable ranges of crops to soils a system of land capability classification has been developed. Land is graded according to the intensity of cropping which can be successful and the level of yield which can be obtained. This type of agricultural interpretation of soil maps was developed in the USA by the Soil Conservation Service. It was designed to classify land on the basis

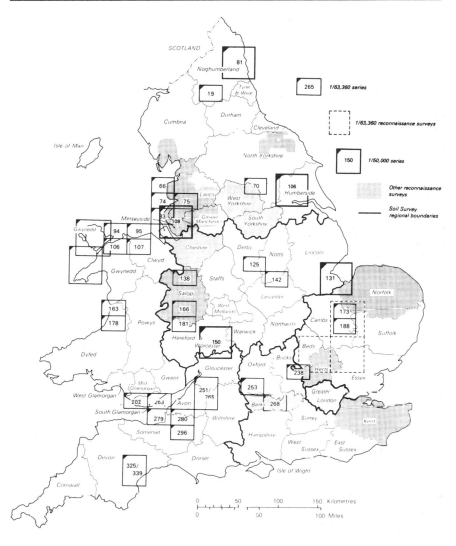

Figure 19.1   Soil Survey of England and Wales

of its suitability for cropping and the measures necessary to
control erosion. A modification of this system is used in recent Soil
Survey reports; for example, the Tideswell Sheet in Derby and the
Honiton Sheet in Devon. Land is graded on a scale of 1–7 on the
basis of the range of crops which are suitable and their levels of
yield. Only physical properties are taken into account, e.g. texture,
slope, drainage and climate. Nutrient contents are not given much

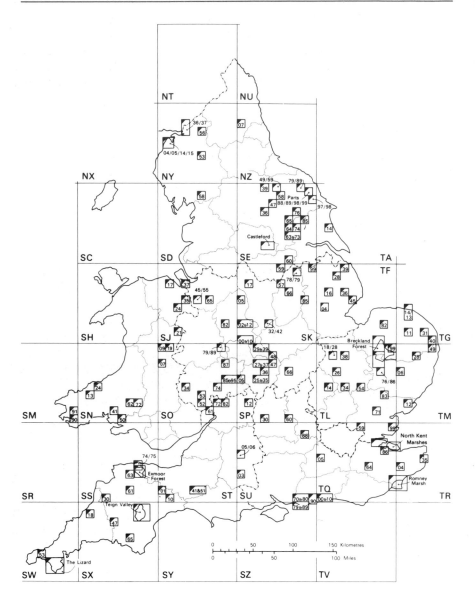

Figure 19.2   Soil Survey of England and Wales

weight in the classification because deficiencies can usually be corrected without much difficulty, in contrast to physical limitations which are more permanent. Land is assessed according to the range of crops which could be successfully grown by a good farmer, which is not necessarily the current land use. Land affected by limitations, e.g. inadequate drainage which can be rectified or reduced at acceptable cost, is graded on the severity of the remaining limitations. Such factors as distance to markets and road access which may change do not influence the grading although they will affect current land use.

Land Capability Classes

Class 1: Land with very minor or no physical limitations to use. Soils in this class are both easy-working and moisture-retentive in a climate suitable to a wide range of crops. They are well drained either naturally or artificially, are level or gently sloping and produce good yields of a wide range of crops. Most Class 1 soils are deep fine sandy loams or peats.

Plate 19.1   Cultivation problems are severe on this poorly structured heavy silt in Holland, Lincolnshire   (Crown copyright)

Class 2: Land with minor limitations which reduce the range of crops that can be grown or the yield which can be obtained. Land in this class includes easy-working land with slight limitations due to proneness to drought, slope, stoniness and adverse drainage or climate. It also includes soils whose range of cropping is somewhat limited by difficulties due to a heavy or unstable topsoil.

Class 3: Land with moderate limitations which restrict the range of crops that can be grown and/or need careful management. For example, this includes light droughty land and also heavy land on which the main arable crops are cereals.

Class 4: Land with moderately severe limitations which restrict the range of crops that can be grown and/or need very careful management. Land in this class includes, for example, very droughty land only capable of producing poor yields without irrigation, and also very difficult heavy or inadequately drained land suited to grass with only an occasional arable crop. Land subject to occasional damaging floods is also in Class 4.

Class 5: Land with severe limitations which restrict agricultural use to pasture.

Class 6: Land with very severe limitations which restrict agricultural use to rough grazing.

Class 7: Land with extremely severe limitations and of little or no agricultural use.

Land Capability and Land Value

The land capability classification will obviously bear some relationship to the monetary value of land, but not a close one. The value of agricultural land depends on other factors as well, in particular proximity to markets or food-processing factories. Class 1 land near to urban areas is likely to be more sought after for vegetable growing, with a price accordingly higher, than land of similar quality with no ready market. Proximity to markets can often be an overriding factor and may make Class 2 or even Class 3 land more desirable and hence more expensive than better land.

Land Capability Maps

Land capability maps indicate capability classes on the scale of 1–7 and also the types of limitation which result in their classification. As an example, land graded as Class 2, because of slight proneness to drought, would have the symbol '2s' on the map which is known as its 'sub-class', the letter 's' indicates a soil limitation. Other soil limitations can be stoniness, shallowness or heaviness.

Land graded as Class 3 because it is steeply sloping has the subclass '3g', where 'g' indicates limitations due to soil gradient. Other symbols used to denote limitations are 'w', which stands for 'wetness' due to inadequate drainage or flooding, 'c' for climatic limitations usually due to altitude, and 'e' denoting susceptibility to erosion. More than one limitation may be recorded in the subclass. Class 3 land so graded because it is heavy and inadequately drained is mapped as subclass 3sw because of both soil and wetness limitations.

From a land capability map it is possible to obtain at a glance a considerable amount of information about the potential of a particular field. Such maps are a valuable aid to farm planning, with a view to growing only crops which are suited to the soils available. The capability classification may indicate that intensification of arable cropping by growing more demanding crops may be practicable or it may indicate the reverse. The dangers of pushing land beyond its capability were highlighted by the wet 1968–9 season in parts of the east Midlands. Inadequate drainage and compacted soil resulted in many crop failures on land in continuous arable which was more suited to ley arable farming.

CLASSIFICATION OF SOME IMPORTANT SOIL SERIES

Hamble Series Class 1

This soil is a deep, well-drained, very fine sandy silt loam (brickearth) and occurs on wind-blown 'loess' on level or gently sloping land. It is found in parts of south-east England including Kent, Hampshire and Essex. Its fine sandy nature gives it a high available water capacity and easy working properties. It is capable of producing consistently high yields of most agricultural and horticultural crops.

Romney Series Class 1

This soil is also a deep, fine sandy silt loam found in silt areas of the Romney Marsh, Cambridgeshire, Norfolk and Lincolnshire. In addition to having a high available water capacity, roots can often reach ground water. It is very productive of a wide range of agricultural and horticultural crops. As it tends to be in exposed positions, it is less suitable for tree fruits.

Adventurers Series Class 1e

Soils of this series are found mainly in the Fen area of East Anglia and consist of peat or peaty loam at least 600 mm in depth. Although the natural drainage was poor, they are now well drained artificially with dyke and pump systems. They are easy working and have a very high available water capacity, so they are unaffected by drought. They produce excellent yields of a wide range of agricultural and vegetable crops. Although susceptible to wind erosion, which occasionally calls for redrilling, good yields are obtained even in years when a blow occurs.

Bromyard Series Class 2s
This red-coloured soil is widespread in the west Midlands, particularly Herefordshire and the Southwest. It has a high available water capacity and so is not droughty, but it is graded Class 2 because of its rather difficult texture, which is a silty clay loam over clay. Although it is well drained, this drainage is rather slow and structural damage is commonly caused by working the soil when wet.

Because of its silty nature it readily slakes and caps. Although a good soil for fruit and hops, it is not very suitable for root crops or intensive vegetable cropping. It is a very good cereals soil.

Hanslope Series Class 2sw
A widespread soil on chalky boulder clay in East Anglia, particularly Cambridgeshire, Essex and west Suffolk, and the east Midlands, particularly Nottinghamshire, Leicestershire and Northamptonshire. It is a deep, very well structured calcareous clay. When drained with a tile and mole system, it becomes virtually well drained. Although heavy it readily weathers to a good tilth. It is an excellent cereal soil and also suitable for some fruit crops. It is capable of growing good crops of roots but structural damage during harvest often adversely affects the next crop.

Wantage Series Class 2s
The Wantage series is widespread on the lower slopes of the chalk areas of southern England and East Anglia, particularly on chalk marls. It is a very good arable soil, but its high chalk content makes it quite unsuitable for fruit. It is slightly affected by drought and tends to be sticky when wet.

Newport Series—Sandy Loam Phase Class 2s
This widespread soil is found in the Midlands and the North, including north Shropshire, Staffordshire, Nottinghamshire (around ADAS Gleadthorpe) and Yorkshire. In East Anglia it was formerly mapped as Freckenham series. It is a well drained and deep, easy working sandy loam, usually slightly stony. It is suitable for a wide range of crops but is slightly droughty.

Newport Series—Loamy Sand Phase Class 3s
This soil is found in the same areas as the sandy loam phase, but is

much more droughty as the subsoil is usually sand and often stony. Without irrigation its productivity is much lower and it is not suitable for vegetable growing apart from carrots.

Evesham Series Class 3sw
This lime-rich soil on Lias clay is important in the east and west Midlands, particularly Warwickshire, and also in Gloucestershire and Somerset. Although well structured and a comparatively easy clay to manage when drained, it is very heavy and suitable mainly for cereals and grass.

Salop Series Class 3sw
This poorly drained soil is a clay loam or fine sandy loam over clay loam formed on Triassic boulder clay; it is widespread in Shropshire, Lancashire and Cheshire and also occurs in Derbyshire, Staffordshire and Nottinghamshire. Without drainage it is suitable only for grass, but when well drained with tiles and moles it is suitable for ley arable farming. It is, however, a difficult soil to manage in arable crops and cultivation pans are readily formed.

Worcester Series Class 3sw
This widespread Keuper Marl soil is found in the east and west Midlands and in the Southwest. It consists of red silty clay loam over silty clay above the marl, which is relatively permeable. Water movement through the silty clay layer is slow and often results in structural damage to the topsoil by working it when wet. With regular subsoiling, which is required to improve the drainage, tile systems are often not needed. It is mainly suitable for cereals and grass.

Icknield Series Class 3s
The Icknield series is a light, very shallow soil over chalk and is common on the upper slopes of the chalklands of southern England and East Anglia. These soils are very well structured and easily cultivated but are moderately susceptible to drought. The main arable crops grown on this soil are spring cereals which can suffer from copper deficiency if the organic matter content is higher than about 7 per cent, although this does not occur in East Anglia.

Sherborne Series Class 3s
The Sherborne series is a shallow reddish-brown soil over limestone. It varies in depth and stoniness and so is prone to drought

in patches. In very dry years striking drought patterns show up in cereals associated with variable depths of rooting. It is important in the Southwest on the Cotswold limestones and also occurs in Northamptonshire around Wansford, near Peterborough, and on the Lincolnshire Cliff. It is usually a clay loam in texture and not difficult to manage, but yields are only moderate in dry years. In the Southwest it mainly supports grass and cereals, but in the east Midlands root crops are grown as well.

Tedburn, Coalpit Heath and Dale Series Class 4ws

The Tedburn series occurs on Culm shales in Devon and strongly resembles the Coalpit Heath and Dale series which occur on coal measures in Somerset and Derbyshire respectively. They consist of poorly drained very heavy clay or silty clay which is unstable and very difficult to drain effectively. They are suitable for only grass with an occasional arable crop.

Fladbury Series Class 4ws

This poorly drained clay occurs in river valleys in many parts of the country. It is subject to flooding and is slow to drain in the spring. It is best suited to pasture, but occasional spring crops can be grown. It cannot be very effectively drained since flooding leads to collapse of moles.

Freckenham Series Class 4s

This coarse sandy soil is common in the Breckland of East Anglia. It is an extremely droughty soil with a loamy sand topsoil over coarse sand or gravel. In some parts of East Anglia, e.g. east Norfolk, the sand in soils of this series is less coarse, so they are less droughty and are graded 3s. It is very deficient in nutrients, particularly potash, magnesium, boron, copper and manganese. Frequent liming is needed. Sugar beet and spring barley are the main crops, but yields vary greatly with season. Docking disorder of sugar beet resulting in stunted and fangy deformed roots is common.

Classes 5–7

Class 5 land is unfit for arable crops, due to very high rainfall, severe climate, poor drainage or steep slopes, but it is suitable for pasture and can be improved by drainage or use of fertilisers, etc. It is often suitable for forestry.

Class 6 land is only suitable for rough grazing which cannot be improved. Its limitations due to very poor drainage, very steep slopes, very severe climate, shallowness or presence of boulders preclude the possibility of improvement by fertiliser use or drainage. It may be suitable for forestry.

Class 7 land is of little or no use in agriculture or forestry and is mainly devoted to recreation. It includes boggy soils, rocky or boulder-strewn soils, bare rock and extremely steep slopes usually with a very severe climate.

IMPROVEMENT OF LAND CAPABILITY

Apart from the obvious things to do, such as improvement of drainage and breaking of pans, is there any way a farmer on poor land can fundamentally improve it? The practice of claying or marling of droughty sands can significantly improve the available water capacity of the topsoil and prevent susceptibility to wind erosion. A considerable amount of land has been treated in this way in the past, but it is not now economic unless clay or marl is available on the same farm. At least 400 tonnes per hectare are required.

Another possible way to improve a droughty soil is to apply pulverised fuel ash from coal-burning power stations, which consists mainly of silt-sized particles and is highly moisture retentive. However, the supplying power station would have to be very close to the land to be treated.

The opposite process—application of sand or ash to lighten heavy soils—is even less practicable. A clay loam soil contains about 40 per cent of clay, and to convert this into an easy-working soil it would be necessary to reduce the clay content by more than one-half, which would require over 2,500 tonnes of sand or ash per hectare. Reducing the clay content of the topsoil by less than a half might reduce the soil's stability and make it even more difficult to manage.

APPENDICES

APPENDIX 1

Soil pH Indicator

Reagents

Methyl red	0.4 g (Hopkin and Williams code No. 5786)
Bromothymol blue	0.8 g (Hopkin and Williams code No. 2376)
Isopropyl alcohol technical 99%	800 ml (Methylating Co. Ltd)
Saturated lime-water	80 ml

Distilled water to make the final volume, 4 litres.

Preparation

Grind and sieve 0.4 g of methyl red through a 70 I.M.M. sieve. Dissolve in 400 ml of isopropyl alcohol.

Dissolve 0.8 gm of bromothymol blue in 400 ml of isopropyl alcohol. Mix the two indicator solutions together, and add to about 3 litres of distilled water.

Add a saturated solution of lime water until the colour becomes green (pH about 7.3), and dilute to 4 litres with distilled water.

For use, transfer to 100 ml polythene bottles, or to glass bottles that release alkali only very slowly from the glass surface.

APPENDIX 2

Rapid Test for Salt in Irrigation Water for Field Crops

Growers and farmers who irrigate crops in areas where water is subject to contamination from salt, which fluctuates rapidly in amount from day to day, require frequent checks on levels of chloride in their water supply. As the analysis is required daily or sometimes even at shorter intervals of time, it is impossible to provide a laboratory analytical service to deal with this. On-the-spot advice is required.

In this system a simple test enables the farmer or adviser to detect whether the supply of water is safe for a particular irrigation. The test shows whether the sample of water contains more or less than 80 kg sodium chloride for every 25 mm of water applied per hectare. If less, the water is safe for one irrigation of up to 25 mm application per hectare, if more it is unsafe and could damage crops.

Methods
Use a Pyrex test-tube or a ground glass stoppered tube, approximately 150 mm in length and 12 mm in diameter calibrated at 10 ml (bottom mark) and 12 ml (top mark). Add the test-water up to the bottom mark followed by 3 drops of *indicator*. Fit a rubber stopper and mix the solution by shaking. Remove the stopper and add sufficient *mercuric nitrate reagent* to bring the level of the solution to the top mark. Fit the stopper and shake to mix the solution. If a violet colour develops the water is safe for irrigation for most agricultural crops. In the absence of a colour development, the water is unsafe.

Reagents
(a) Mercuric Nitrate: $Hg(NO_3)_2H_2O$
Dissolve 11.8 g mercuric nitrate in 50 ml (approx.) distilled water to which has been added 15 ml conc. HNO_3, and make up to 1 litre with distilled water.
(b) Indicator: Diphenyl carbazone
Dissolve 0.5 g diphenyl carbazone in 100 ml ethyl alcohol.
Note: Mercuric nitrate is a poison and the normal precautions should be taken when handling and using it.

Taken from Eastern region ADAS circular 67/13

APPENDIX 3

Metric Conversion Table

Weight	1 tonne	=	1000 kilograms (kg)
		=	0.984 ton
		=	19.7 cwt
	1 kg	=	2.20 lb
Length	1 kilometre (km)	=	1.61 mile
	1 metre (m)	=	1.09 yard
	1 millimetre (mm)	=	0.039 inch
Area	1 hectare (ha)	=	2.47 acres
Volume	1 litre (1)	=	0.22 gallon
		=	0.035 cu ft
Yield	1 tonne/ha	=	0.398 ton/acre
		=	7.96 cwt/acre
Rates	1 kg/ha	=	0.89 lb/acre
		=	0.08 unit/acre

INDEX

275